Genetics and Breeding of Plants

Genetics and Breeding of Plants

Ava Gibson

MURPHY & MOORE
www.murphy-moorepublishing.com

Published by Murphy & Moore Publishing,
1 Rockefeller Plaza,
New York City, NY 10020, USA
www.murphy-moorepublishing.com

Genetics and Breeding of Plants
Ava Gibson

International Standard Book Number: 978-1-63987-255-8 (Hardback)

Cataloging-in-Publication Data

Genetics and breeding of plants / Ava Gibson.
p. cm.
Includes bibliographical references and index.
ISBN 978-1-63987-255-8
1. Plant genetics. 2. Plant breeding. 3. Genetics. 4. Breeding.
5. Agriculture. I. Gibson, Ava.

SB123 .G46 2022
631.53--dc23

TABLE OF CONTENTS

PREFACE

The branch of biology which focuses on genetic variations and heredity in case of plants is called plant genetics. It is used to modify the genetic structure of staple crops to increase yields, provide resistance against herbicides, improve disease resistance, and increase their nutritional value. Plant breeding refers to the science of changing plant traits in order to obtain desired characteristics. The key objectives of plant breeding are to improve biotic and abiotic stress tolerance, biomass yields, and end-use quality characteristics. Some of the important modification techniques studied under the discipline of plant breeding are marker assisted selection, reverse breeding and double haploids, and genetic modification. This book is compiled in such a manner, that it will provide in-depth knowledge about the theory and practice of plant breeding and genetics. Some of the diverse topics covered herein address the varied branches that fall under this category. This book is an essential guide for both academicians and those who wish to pursue this discipline further.

Given below is the chapter wise description of the book:

Chapter 1- Plant genetics is the study of genes, genetic variations and heredity in plants. It also deals with the mechanisms of hereditary transmission and variation of inherited characteristics. The scientific process of developing new plant varieties by careful selection and crossing of plants with desired traits is known as plant breeding. This chapter has been carefully written to provide an easy understanding of plant genetics and breeding.

Chapter 2- Plant reproduction is the process by which plants generate new individuals or offspring. This can be accomplished through either sexual or asexual means. Sexual reproduction involves the fusion of gametes while asexual reproduction may occur through budding, fragmentation or spore formation. This chapter closely examines these types of plant reproduction for a better understanding of the topic.

Chapter 3- Plant molecular breeding may be broadly defined as the use of genetic manipulation at DNA molecular levels in order to alter or improve certain desirable characteristics in plants. Molecular markers are specific DNA fragments in a genome which are used in marker-assisted breeding to achieve desired changes. All aspects of molecular breeding in plants are discussed in detail in this chapter.

Chapter 4- Selection is an important process as it defines the characteristics of the offspring. Various selection methods are employed based on desired outcome and plant characteristics. Breeding is the next step and several breeding methods are available based on the plant's mode of reproduction and pollination. This chapter deals with the selection and breeding methods for self-pollinated, cross-pollinated, hybrid cultivars and clonally propagated plants in a comprehensive manner.

Chapter 5- Genetic modification in plants has come a long way and has resulted in many genetically modified crops being produced and cultivated commercially. These crops have better traits and qualities which make them much more viable for cultivation, consumption and commercial purposes. The topic of genetically modified plants is covered in detail in this chapter.

At the end, I would like to thank all those who dedicated their time and efforts for the successful completion of this book. I also wish to convey my gratitude towards my friends and family who supported me at every step.

Ava Gibson

Chapter 1

Introduction to Plant Genetics and Breeding

Plant genetics is the study of genes, genetic variations and heredity in plants. It also deals with the mechanisms of hereditary transmission and variation of inherited characteristics. The scientific process of developing new plant varieties by careful selection and crossing of plants with desired traits is known as plant breeding. This chapter has been carefully written to provide an easy understanding of plant genetics and breeding.

Plant Genetics

Genetics is a branch of biology that deals with heredity of genetic properties and variation of organisms. Gregor Johann Mendel is called the father of genetics because he discovered how traits were inherited. Previously, heredity was regarded as a "blending" theory and offspring were a blended mix of both parents. Mendel was an Austrian Monk who attended the University of Vienna and studied botany. He worked with peas for eight years and looked at seven different traits describing seed texture and color, flower color, pod texture and color, flower position and stem length.

Figure: Picture from "Mendel's experiments.

Mendel was the first botanist to use math to explain his results quantitatively. He discovered principles of heredity that are the foundation for genetics today. Genetics is a powerful tool used in many different fields, including Plant Science. Plant scientists use genetics to create new strains of plants with traits of interest (e.g. flower color or seed Picture from "Mendel's experiments). Plant scientists also change the expression of genes to create "mutants" which alter biological processes and enable discovery of genes control a biological process. Understanding genetics is extremely important to understanding plants and plant function. DNA (deoxyribonucleic acid) contains the genetic material that makes up an organism. DNA is made up of two chains with different arrangements of nucleotides. How nucleotides are arranged determines the genetic information. Genes are a section of the DNA sequence that describes various traits found in an organism. All genes are located on chromosomes, which contain the organized genetic material.

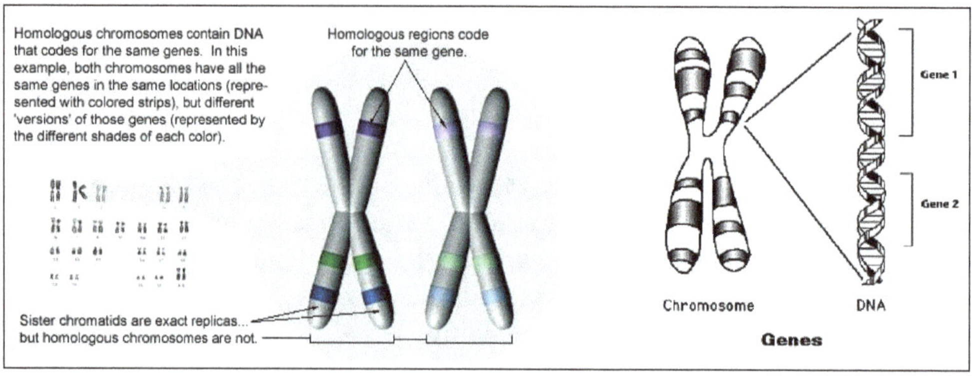

Alleles are versions of genes that occupy the same position on homologous chromosomes and express variations of a trait. When an organism has identical alleles for a specific trait is called homozygous (ex: TT). An organism having two different alleles for a particular trait is called heterozygous (ex: Tt). The complete genetic make-up of an organism is called a genotype, whereas the phenotype is the observable physical appearance of an organism. If an allele is dominant, then that allele of the gene masks or suppresses the expression of an alternate allele. If an allele is recessive, then that allele is masked by a dominant allele and does not appear in the heterozygous condition but rather in the homozygous condition.

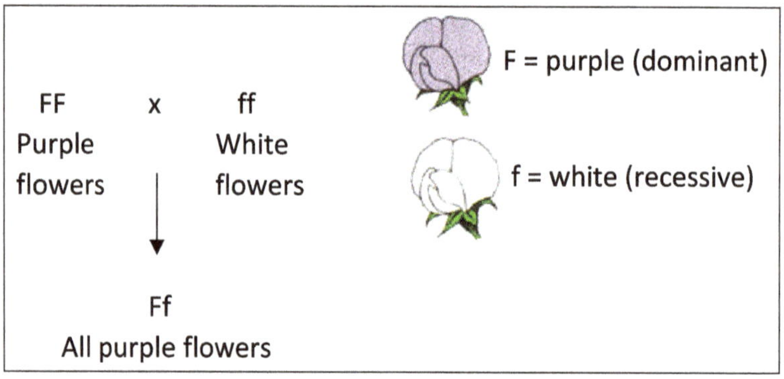

Figure: Example with flower color: P = Parental generation Parental genotype; F1 generation: F1 phenotype.

We can use a Punnett square to understand the genotypes and phenotypes of the offspring between two parents.

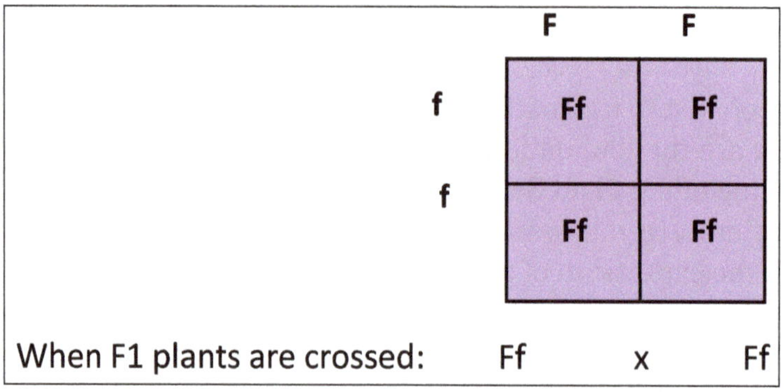

Figure: Genotypes: 4 Ff – purple flowers Phenotypes: 4 purple flowers.

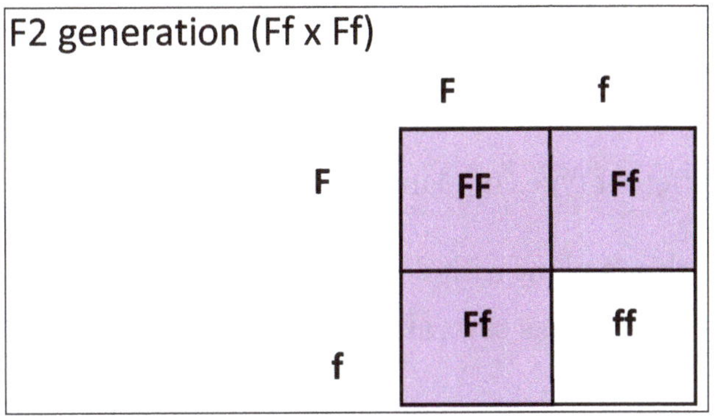

Figure: Genotypes: 1 FF – purple 2 Ff – purple 1 ff – white Phenotypes: 3 purple 1 white.

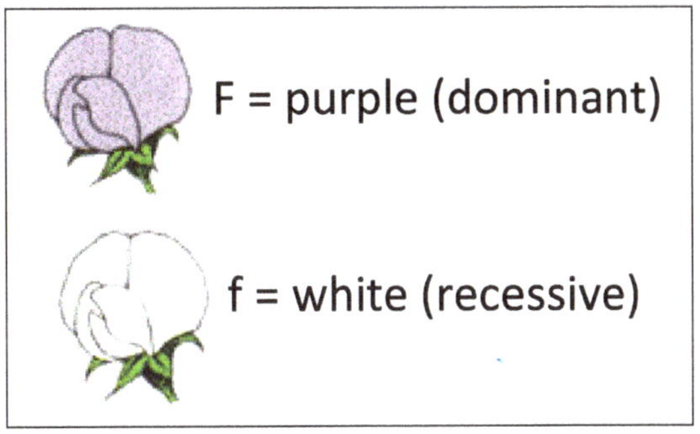

We ultimately end up with a genotypic ratio of 1 FF : 2Ff : 1 ff and the phenotypic ratio is 3 purple flowers : 1 white flower.

Using Punnett squares, we can predict the phenotype of the offspring. Plant breeders use genetic information to make combinations to improve plants. They select plants with desirable traits to be parents and cross them to create progeny with improved characteristics. Norman Borlaug is the most well-known plant breeder who won the 1970 Nobel Peace Prize for his contribution through breeding of high-yielding semi-dwarf varieties of wheat and maize. He is credited for saving 1 billion people worldwide.

Plant Breeding

Plant breeding is the purposeful manipulation of qualities in plants to create new varieties with a set of desired characteristics. Plants with higher qualities are selected by and crossed to obtain plants with desired quality. This results in a plant population with improved and desired traits.

Following are the major objectives of plant breeding:

- To increase the crop yield.
- To raise plants with desired characteristics.
- To develop a disease-resistant crop.

- To develop plants that can tolerate extreme environmental stress.

Types of Plant Breeding

There are the following different types of plant breeding:

- Backcrossing: In this, a plant with desired traits is crossed with a plant that does not have the desired traits but has several other traits.
- Inbreeding: In this, self-fertilization occurs. The progeny produced is the same generation after generation. This helps to preserve the original traits.
- Hybrid Breeding: In this, two different breeds are crossed to produce the offspring that is more productive than the parents.
- Mutation Breeding: The mutations in plant genes result in new varieties. Mutations can also be induced in plants by exposing them to chemicals and radiation.
- Genetic Engineering: Genetic engineering helps in producing crops with desirable traits by inserting the gene of interest within the crop DNA. Such crops are known as genetically modified crops. E.g., Bt crops

Plant Breeding Steps

The plant breeding methods have undergone multiple amendments since it was started from 9000 – 11000 years ago. The modern method of plant breeding is carried out in the following steps:

Collection of Variability

Variability forms the root of all breeding practices. The first step involves the collection of plants or seeds for all possible alleles for all genes in a given crop, which is known as germplasm. This collection even involves wild varieties and relatives of the cultivated species.

Evaluation and Selection of Parent Plants

The Germplasm is evaluated for selection of parent plants with desired characteristics. The combination of these characteristics is expected in the hybrid progeny.

For example, a plant crop with high protein content can be selected to be crossed with a plant with higher disease resistance.

Cross-hybridization Among Selected Parents

In the third step, the parents are cross-hybridized to develop pure lines of progenies. This is a tedious and time-consuming practice carried out in the conventional fashion of introducing pollen from one plant to the stigma of the other. Despite the labour involved, only one or two progenies in a few hundred show the desired combination of characteristics.

Selection and Testing of Superior Recombinants

The developed progenies are then evaluated and those with the desired combination of characteristics are self-pollinated to attain homozygosity.

Testing, Release and Commercialization of New Cultivars

The new lines are grown in research fields where they are tested for their agronomic traits of quality, yield, resistance to diseases, etc. This is followed by the cultivation of these crops in farmers' fields at different locations in the country which represent the different agro climatic zones. With successful results, the crops are released commercially for public consumption.

Genetic Engineering

Genetic engineering is a new type of genetic modification. It is the purposeful addition of a foreign gene or genes to the genome of an organism. A gene holds information that will give the organism a trait. Genetic engineering is not bound by the limitations of traditional plant breeding. Genetic engineering physically removes the DNA from one organism and transfers the gene(s) for one or a few traits into another. Since crossing is not necessary, the 'sexual' barrier between species is overcome. Therefore, traits from any living organism can be transferred into a plant. This method is also more specific in that a single trait can be added to a plant.

Step 1: DNA Extraction

The process of genetic engineering requires the successful completion of a series of five steps. DNA extraction is the first step in the genetic engineering process. In order to work with DNA, scientists must extract it from the desired organism. A sample of an organism containing the gene of interest is taken through a series of steps to remove the DNA.

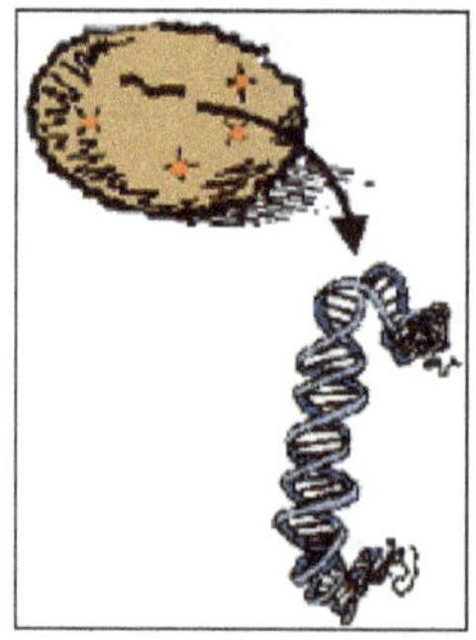

Extracting DNA from and organism.

Step 2: Gene Cloning

Second step of the genetic engineering process is gene cloning. During DNA extraction, the entire DNA from the organism is extracted at once. Scientists use gene cloning to separate the single gene of interest from the rest of the genes extracted and make thousands of copies of it.

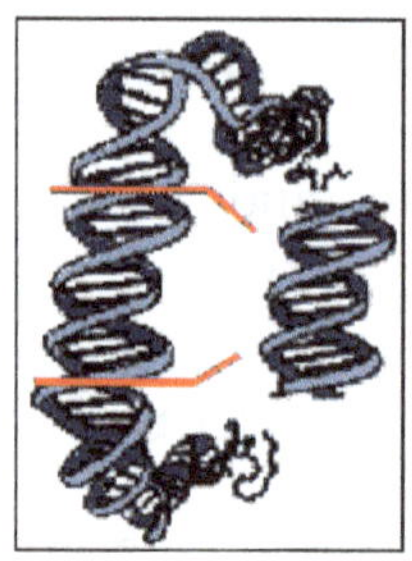

Selecting a gene for cloning·

Step 3: Gene Design

Once a gene has been cloned, genetic engineers begin the third step, designing the gene to work once inside a different organism. This is done in a test tube by cutting the gene apart with enzymes and replacing gene regions that have been separated.

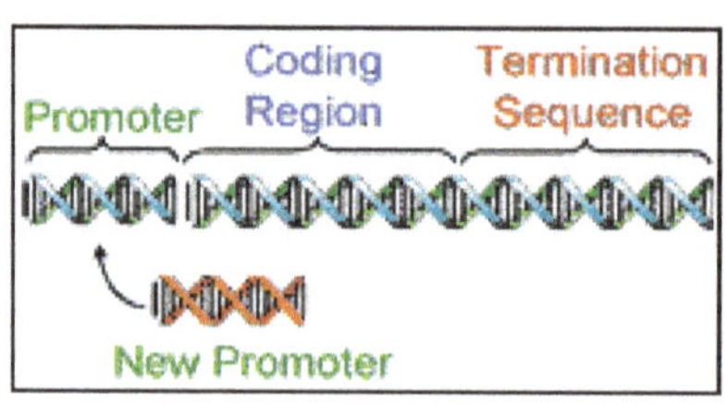

Replacing an existing promoter with a new promoter.

Step 4: Transformation

The modified gene is now ready for the fourth step in the process, transformation or gene insertion.

Since plants have millions of cells, it would be impossible to insert a copy of the transgene into every cell. Therefore, tissue culture is used to propagate masses of undifferentiated plant cells called callus. These are the cells to which the new transgene will be added.

The new gene is inserted into some of the cells using various techniques. Some of the more common methods include the gene gun, agrobacterium, microfibers, and electroporation. The main goal of each of these methods is to transport the new gene(s) and deliver them into the nucleus of a cell without killing it. Transformed plant cells are then regenerated into transgenic plants. The transgenic plants are grown to maturity in greenhouses and the seed they produce, which has inherited the transgene, is collected. The genetic engineer's job is now complete. He/she will hand the transgenic seeds over to a plant breeder who is responsible for the final step.

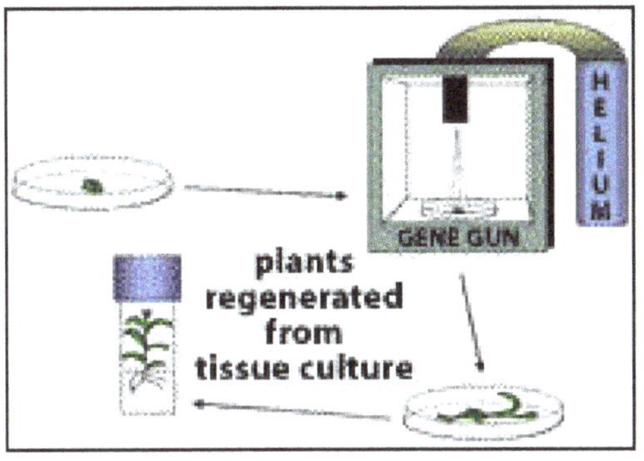

Using the gene gun method to transform plant cells.

Step 5: Backcross Breeding

The fifth and final part of producing a genetically engineered crop is backcross breeding. Transgenic plants are crossed with elite breeding lines using traditional plant breeding methods to combine the desired traits of elite parents and the transgene into a single line. The offspring are repeatedly crossed back to the elite line to obtain a high yielding transgenic line. The result will be a plant with a yield potential close to current hybrids that expresses the trait encoded by the new transgene.

Backcross breeding.

The Process of Plant Genetic Engineering

The entire genetic engineering process is basically the same for any plant. The length of time required to complete all five steps from start to finish varies depending upon the gene, crop species, available resources and regulatory approval. It can take anywhere from 6-15+ years before a new transgenic hybrid is ready for release to be grown in production fields.

Plant Genetic Modification

A genetically heterogeneous population of plants is inspected, and "superior" individuals—plants with the most desired traits, such as improved palatability and yield—are selected for continued propagation. The others are eaten or discarded. The seeds from the superior plants are sown to produce a new generation of plants, all or most of which will carry and express the desired traits. Over a period of several years, these plants or their seeds are saved and replanted, which increases the population of superior plants and shifts the genetic population so that it is dominated by the superior genotype. This very old method of breeding has been enhanced with modern technology.

An example of modern methods of simple selection is marker-assisted selection, which uses molecular analysis to detect plants likely to express desired features, such as disease resistance to one or more specific pathogens in a population. Successfully applying marker-assisted selection allows a faster, more efficient mechanism for identifying candidate individuals that may have "superior traits."

Superior traits are those considered beneficial to humans, as well as to domesticated animals that consume a plant-based diet; they are not necessarily beneficial to the plant in an ecological or evolutionary context. Often traits considered beneficial to breeders are detrimental to the plant from the standpoint of environmental fitness. For example, the reduction of unpalatable chemicals in a plant makes it more appealing to human consumers but may also attract more feeding by insects and other pests, making it less likely to survive in an unmanaged environment. As a result, cultivated crop varieties rarely establish populations in the wild when they escape from the farm. Conversely, some traits that enhance a plant's resistance to disease may also be harmful to humans.

Crossing

Crossing occurs when a plant breeder takes pollen from one plant and brushes it onto the pistil of a sexually compatible plant, producing a hybrid that carries genes from both parents. When the hybrid progeny reaches flowering maturity, it also may be used as a parent.

Plant breeders usually want to combine the useful features of two plants. For example, they might add a disease-resistance gene from one plant to another that is high-yielding but disease-susceptible, while leaving behind any undesirable genetic traits of the disease-resistant plant, such as poor fertility and seed yield, susceptibility to insects or other diseases, or the production of antinutritional metabolites.

Because of the random nature of recombining genes and traits in crossed plants, breeders usually have to make hundreds or thousands of hybrid progeny to create and identify those few that possess useful features with a minimum of undesirable features. For example, the majority of progeny may show the desired disease resistance, but unwanted genetic features of the disease-resistant parent may also be present in some. Crossing is still the mainstay of modern plant breeding, but many other techniques have been added to the breeders' tool kit.

Interspecies Crossing

Interspecies crossing can take place through various means. Closely related species, such as cultivated oat (Avena sativa) and its weedy relative wild oat (Avena fatua), may cross-pollinate for exchange of genetic information, although this is not generally the case. Genes from one species also can naturally integrate into the genomes of more distant relatives under certain conditions. Some food plants can carry genes that originate in different species, transferred both by nature and by human intervention. For example, common wheat varieties carry genes from rye. A common potato, Solanum tuberosum, can cross with relatives of other species, such as S. acaule or S. chacoense.

Chromosome engineering is the term given to nonrecombinant deoxyribonucleic acid (rDNA) cytogenetic manipulations, in which portions of chromosomes from near or distant species are recombined through a natural process called chromosomal translocation. Sears pioneered the human exploitation of this process, which proved valuable for transferring traits that were otherwise unattainable, such as pest or disease resistance, into crop species. However, because transferring large segments of chromosomes also transferred a number of neutral or detrimental genes, the utility of this technique was limited.

Recent refinements allow plant breeders to restrict the transferred genetic material, focusing more on the gene of interest. As a result, chromosome engineering is becoming more competitive with rDNA technology in its ability to transfer relatively small pieces of DNA. Several crop species, such as corn, soybean, rice, barley, and potato, have been improved using chromosome engineering.

Embryo Rescue

Sometimes human technical intervention is required to complete an interspecies gene transfer. Some plants will cross-pollinate and the resulting fertilized hybrid embryo develops but is unable to mature and sprout. Modern plant breeders work around this problem by pollinating naturally and then removing the plant embryo before it stops growing, placing it in a tissue-culture environment where it can complete its development. Such embryo rescue is not considered genetic engineering, and it is not commonly used to derive new varieties directly, but it is used instead as an intermediary step in transferring genes from distant, sexually incompatible relatives through intermediate, partially compatible relatives of both the donor and recipient species.

Somatic Hybridization

Recent advances in tissue-culture technologies have provided new opportunities for recombining genes from different plant sources. In somatic hybridization, a process also known as cell fusion, cells growing in a culture medium is stripped of their protective walls, usually using pectinase, cellulase, and hemicellulase enzymes. These stripped cells, called protoplasts, are pooled from different sources and, through the use of varied techniques such as electrical shock, are fused with one another.

When two protoplasts fuse, the resulting somatic hybrid contains the genetic material from both plant sources. This method overcomes physical barriers to pollen-based hybridization, but not basic chromosomal incompatibilities. If the somatic hybrid is compatible and healthy, it may grow a new cell wall, begin mitotic divisions, and ultimately grow into a hybrid plant that carries genetic features of both parents. While protoplast fusions are easily accomplished, as almost all plants (and animals) have cells suitable for this process, relatively few are capable of regenerating a whole organism, and fewer still are capable of sexual reproduction. This non-genetic engineering technique is not common in plant breeding as the resulting range of successful, fertile hybrids has not extended much beyond what is possible using other conventional technologies.

Somaclonal Variation

Somaclonal variation is the name given to spontaneous mutations that occur when plant cells are grown in vitro. For many years plants regenerated from tis-sue culture sometimes had novel features. It was not until the 1980s that two Australian scientists thought this phenomenon might provide a new source of genetic variability, and that some of the variant plants might carry attributes of value to plant breeders.

Through the 1980s plant breeders around the world grew plants in vitro and scored regenerants for potentially valuable variants in a range of different crops. New varieties of several crops, such as flax, were developed and commercially released. Molecular analyses of these new varieties were not required by regulators at that time, nor were they conducted by developers to ascertain the nature of the underlying genetic changes driving the variant features. Somaclonal variation is still used by some breeders, particularly in developing countries, but this non-genetic engineering technique has largely been supplanted by more predictable genetic engineering technologies.

Mutation Breeding: Induced Chemical and X-ray Mutagenesis

Mutation breeding involves exposing plants or seeds to mutagenic agents (e.g., ionizing radiation) or chemical mutagens (e.g., ethyl methanesulfonate) to induce random changes in the DNA sequence. The breeder can adjust the dose of the mutagen so that it is enough to result in some mutations, but not enough to be lethal. Typically a large number of plants or seeds are mutagenized, grown to reproductive maturity, and progeny are derived. The progeny are assessed for phenotypic expression of potentially valuable new traits.

As with somaclonal variation, the vast majority of mutations resulting from this technique are deleterious, and only chance determines if any genetic changes useful to humans will appear. Other than through varying the dosage, there is no means to control the effects of the mutagen or to target particular genes or traits. The mutagenic effects appear to be random throughout the genome and, even if a useful mutation occurs in a particular plant, deleterious mutations also will likely occur. Once a useful mutation is identified, breeders work to reduce the deleterious mutations or other undesirable features of the mutated plant. Nevertheless, crops derived from mutation breeding still are likely to carry DNA alterations beyond the specific mutation that provided the superior trait.

Induced-mutation crops in most countries are not regulated for food or environmental safety, and breeders generally do not conduct molecular genetic analyses on such crops to characterize the mutations or determine their extent. Consequently, it is almost certain that mutations other than those resulting in identified useful traits also occur and may not be obvious, remaining uncharacterized with unknown effects.

Cell Selection

Several commercial crop varieties have been developed using cell selection, including varieties of soybeans, canola, and flax. This process involves isolating a population of cells from a so-called "elite plant" with superior agricultural characteristics. The cells are then excised and grown in culture. Initially the population is genetically homogeneous, but changes can occur spontaneously (as in somaclonal variation) or be induced using mutagenic agents. Cells with a desired phenotypic variation may be selected and regenerated into a whole plant. For example, adding a suitable amount of the appropriate herbicide to the culture medium may identify cells expressing a novel variant phenotype of herbicide resistance. In theory, all of the normal, susceptible cells will succumb to the herbicide, but a newly resistant cell will survive and perhaps even continue to grow. An herbicide-resistant cell and its derived progeny cell line thus can be selected and regenerated into a whole plant, which is then tested to ensure that the phenotypic trait is stable and results from a heritable genetic alteration. In practice, many factors influence the success of the selection procedure, and the desired trait must have a biochemical basis that lends itself to selection in vitro and at a cellular level.

Breeders cannot select for increased yield in cell cultures because the cellular mechanism for this trait is not known. The advantage of cell selection over conventional breeding is the ability to inexpensively screen large numbers of cells in a petri dish in a short time instead of breeding a similar number of plants in an expensive, large field trial conducted over an entire growing season.

Like somaclonal variation, cell selection has largely been superceded by recombinant technologies because of their greater precision, higher rates of success, and fewer undocumented mutations.

Variety of Genetic Engineering Techniques

Defines genetic engineering specifically as one type of genetic modification that involves an intended targeted change in a plant or animal gene sequence to effect a specific result through the use of rDNA technology. A variety of genetic engineering techniques are described in the following text.

Microbial Vectors

Agrobacterium *tumefaciens* is a naturally occurring soil microbe best known for causing crown gall disease on susceptible plant species. It is an unusual pathogen because when it infects a host, it transfers a portion of its own DNA into the plant cell. The transferred DNA is stably integrated into the plant DNA, and the plant then reads and expresses the transferred genes as if they were its own. The transferred genes direct the production of several substances that mediate the development of a crown gall.

Among these substances is one or more unusual non-protein amino acids, called opine. Opines are translocated throughout the plant, so food developed from crown gall-infected plants will carry these opines. In the early 1980s strains of Agrobacterium were developed that lacked the disease-causing genes but maintained the ability to attach to susceptible plant cells and transfer DNA.

By substituting the DNA of interest for the crown gall disease-causing DNA, scientists derived new strains of Agrobacterium that deliver and stably integrate specific new genetic material into the cells of target plant species. If the transformed cell then is regenerated into a whole fertile plant, all cells in the progeny also carry and may express the inserted genes. Agrobacterium is a naturally occurring genetic engineering agent and is responsible for the majority of GE plants in commercial production.

Micro-projectile Bombardment

Klein and colleagues discovered that naked DNA could be delivered to plant cells by "shooting" them with microscopic pellets to which DNA had been adhered. This is a crude but effective physical method of DNA delivery, especially in species such as corn, rice, and other cereal grains, which Agrobacterium does not naturally transform. Many GE plants in commercial production were initially transformed using micro-projectile delivery.

Electroporation

In electroporation, plant protoplasts take up macromolecules from their surrounding fluid, facilitated by an electrical impulse. Cells growing in a culture medium are stripped of their protective walls, resulting in protoplasts. Supplying known DNA to the protoplast culture medium and then applying the electrical pulse temporarily destabilizes the cell membrane, allowing the DNA to enter the cell. Transformed cells can then regenerate their cell walls and grow to whole, fertile transgenic plants. Electroporation is limited by the poor efficiency of most plant species to regenerate from protoplasts.

Microinjection

DNA can be injected directly into anchored cells. Some proportion of these cells will survive and integrate the injected DNA. However, the process is labor intensive and inefficient compared with other methods.

Transposons/Transposable Elements

The genes of most plant and some animal (e.g., insects and fish) species carry transposons, which are short, naturally occurring pieces of DNA with the ability to move from one location to another in the genome. Barbara McClintock first described such transposable elements in corn plants during the 1950s. Transposons have been investigated extensively in research laboratories, especially to study mutagenesis and the mechanics of DNA recombination. However, they have not yet been harnessed to deliver novel genetic information to improve commercial crops.

Non-transgenic Molecular Methods of Manipulation

Genetic features can be added to plants and animals without inserting them into the recipient organism's native genome. DNA of interest may be delivered to a plant cell, expressing a new protein—and thereby a new trait—without becoming integrated into the host-cell DNA. For example, virus strains may be modified to carry genetic material into a plant cell, replicate, and thrive without integrating into the host genome. Without integration, however, new genetic material may be lost during meiosis, so that seed progeny may not carry or express the new trait.

Many food plants are perennials or are propagated by vegetative means, such as grafting or from cuttings. In these cases the virus and new genes would be maintained in subsequent, nonsexually generated populations. Technically such plants are not products of rDNA because there is no recombination or insertion of introduced DNA into the host genome. Although these plants are not GE, they do carry new DNA and new traits.

Transgenic Plants

Transgenic plants are plants that have had their genomes modified through genetic engineering techniques either by the addition of a foreign gene or removal of a certain detrimental gene. A foreign gene inserted into a plant can be of a different species or even kingdom. The first transgenic plant was developed through the insertion of nptII bacterial antibiotic resistance gene into tobacco. Since then, with the rapid development in plant molecular biology and genetic engineering technology, a wide variety of transgenic plants with important agronomic traits such as pest resistance and drought tolerance have been developed, ranging from dicots to monocots that are amenable to genetic modifications. The main purpose in the production of transgenic plants is to produce crops, which have ideal traits, quality, and high yield. Besides being beneficial to the agriculture sector, the plants are found to be able to act as the factory for pharmaceutical protein production.

Application of Transgenic Plants

Resistance to Biotic or Abiotic Stresses

Biotic stresses occur naturally as a result of stress exerted from other living organism within the same ecosystem. These include bacteria, viruses, herbivores, or native plants. Crop plants are incorporated with disease resistance gene to confer resistance toward these pathogenic diseases that are caused by pest, bacteria, and viruses; this includes tolerance to herbicides. The introduction of genetic modification technology could reduce the usage of expensive pesticides and herbicides in agriculture. The removal of natural pests will lead to a greater yield and better quality of crops. As such, insecticidal toxin genes from a bacterium can be introduced into the plant of interest's genome, thus providing protection to the plant against insect pests. Bacillus thuringiensis (Bt) crops are an example of transgenic plant produced through this method. In addition, virus-resistant plants can be achieved through the introduction of viral coat proteins into plants.

Development of transgenic plants resistant to abiotic stresses is important in this "Global Warming's Terrifying Era". The world climate in the past few decades has changed tremendously culminating in changes to soil composition, humidity, water, sunlight availability, and many other agricultural problems that led to reduction in the crop yield. Hence, genetic engineering technology is needed as a tool to solve these problems by providing the plants with enhanced stress tolerant ability or protection. The manipulation of transcription factors (TFs), late embryogenesis abundant (LEA) proteins, and antioxidant proteins had successfully produced plants tolerant to drought and salinity. Overexpression of the proline biosynthesis enzyme (P5C), which allows the accumulation of osmoprotectant during drought season, provides transgenic plants with osmotic stress resistance.

Improving Crop Yield and Nutritional Value

Malnutrition is a major health concern that is prevalent especially in the underdeveloped and developing countries due to limited access to nutritious food. Genetic engineering of staple crops has become one of the more effective solutions in addressing this problem. To date, a variety of crops had been successfully modified for better yield as well as for higher nutritional value. Biofortification is a technique used in agriculture to increase the nutritional value of crops. A well-known example would be the golden rice, a variety of Oryza sativa, produced to biosynthesize beta-carotene through genetic modification. The golden rice was developed by adding two beta-carotene synthesis genes: phytoene synthase (psy) and lycopene β-cyclase (β-lcy) (originated from Narcissus pseudo narcissus). These genes were driven under the control of the endosperm-specific glutelin promoter together with a bacterial phytoene desaturase (crtI, from Erwinia uredovora).

Transgenic Plants as Bioreactors for Recombinant Proteins

Plants had been used as a biofactory in the production of the first recombinant human protein in 1989. Product yields from recombinant proteins using mammalian expression systems are low and expensive, while bacteria system is incapable of post-translational modification in complex protein formation. Due to this, the production methods had shifted to plant cell systems, which provide cheaper and better alternative sources for recombinant proteins production. The recombinant proteins produced in transgenic plants include antibodies, metabolites or catabolites, proteins, and vaccines. Antibodies and vaccines against gastrointestinal tract diseases, cholera, and malaria are known to be produced in transgenic plants such as potato, banana, algae, and tobacco. An anticancer antibody that recognizes the cells of lung, breast, and colon cancer had also been successfully expressed in rice and wheat seed.

Gene Constructs

A simple functional gene construct consists of a promoter region, gene coding region, and terminator/stop region. In addition, certain gene constructs may contain special sequences such as an enhancer, silencer, or reporter sequences depending on the nature of study. Plant transformation always starts with the transgene construction. Transgene construct generally has similar elements other than the inclusion of the gene of interest and selectable markers. A proper gene construct is crucial for the success of producing ideal transgenic line.

A Typical Plant Gene

A typical plant gene consists of the regulatory and structural genes. Regulatory genes are usually located at the 5′ upstream of a gene, with its own promoter, enhancer, or silencer region. Structural genes, on the other hand, begin with a catabolite activator protein (cap) site, followed by a leader sequence, start codon, exons, introns, terminator, and a polyadenylation site (poly-A tail). These elements are responsible for DNA transcription. The transcribed pre-mRNA then undergoes RNA splicing, producing mature mRNA without the introns (noncoding region). This mature mRNA is delivered to the cytoplasm for translation initiated by the binding of ribosomal subunits to the promoter. Translation then begins at the start codon (ATG), with the ribosome moving downstream to the next codon creating a peptide chain with the help of tRNAs and ends once it reaches the chain terminator (stop codon, TAA/TAG).

Promoters/Enhancers

The promoter region is typically located at the 5′ upstream of a gene. Promoters are known for their function in governing gene expression, likened to an on/off switch. In DNA transcription, the promoter sequence is recognized by transcription factors. These transcription factors bind to the consensus region of the promoter and recruit the RNA polymerase. Formation of the RNA polymerase transcription complex marks the beginning of DNA transcription.

The promoters can be categorized into three main groups: constitutive promoters, tissue-specific promoters, and inducible promoters. The constitutive promoters are active at most of the developmental stages, and they directly participate in maintaining moderate and constant level of gene expression. Tissue-specific promoters provide restricted gene expression to certain tissues or gene expression involves in developmental-specific stages. Gene expressions associated with the inducible promoters are greatly affected by environmental stimuli, which allow for the regulation of genes through external factors. Table below shows selected promoters used in plant transformation.

Table: Examples of promoter used in plant transformation.

Promoter	Source	Activity
CaMV$_{35}$S	Cauliflower mosaic virus	Constitutive
Ubiquitin RUBQ$_1$, RUBQ$_2$ and rubi$_3$	Rice	Constitutive
Ubiquitin Gmubi$_3$	Soybean	Constitutive
SCR, SRK	Brassica rapa	Pollen and stigma specific
E$_{x070}$C$_2$	Arabidopsis	Pollen and stigma specific
LMW Glu, HMW Glu$_{-1}$D$_1$	Wheat	Seed specific
Expansin PcExp$_2$	Sour cherry	Ripened fruits
Potato class I patatin	Potato	Tuber/storage organ specific
NtHSP$_3$A	Tobacco	Stress inducible

Enhancers are short (50–1500 bp) regions in a gene that can be recognized and bound by activator proteins. These proteins, also referred to as transcription factors, bind to the enhancer, forming an enhancer-bound transcription factor complex, which will later on interact with the mediator complex (TFIID) ultimately aiding in the recruitment of RNA polymerase II. The enhancer-bound transcription factor complex forms a loop and toward the intervening sequence and comes in contact with the promoter region, thus increasing the accessibility of the promoter to the transcription proteins. In contrast, silencers function as the direct opposite of enhancers. Silencers are binding sites for transcription factors known as the repressors. These repressors are known to down regulate the transcription of a gene. In plant genetic engineering, suitable promoter and enhancer are chosen based on the intended regulation of gene expression. Gene expression is kept at basal level when the transgene exerts mild toxicity to the target plant. On the other hand, higher gene expression levels facilitate the detection and monitoring of a transgene which may usually be under expressed in nature.

Reporter Genes

Reporter genes are genes attached to the regulatory sequences or to gene of interest to allow for detection of the transgene expression as well as the localization of expressed proteins. Reporter gene sequences encode proteins or products of the protein after being catalyzed for detection through instruments or simple assays. In contrast, selectable marker genes such as antibiotic genes, herbicidal-resistant genes, and anti-metabolic genes confer resistance toward certain chemical agents, which inhibit nontransgenic plant development. The common reporter genes used to monitor plant transgene expression include green fluorescent protein (GFP), chloramphenicol acetyltransferase (CAT), beta-galactosidase (LacZ), luciferase (Luc), and beta-glucuronidase (GUS). These reporter genes allow differentiation between transformed and nontransformed cells and enable detection of transgene localization and regulation of the expressed and tagged protein. Dual reporter systems such as Luc/Luc and GUS/Luc are also available for better detection in distinguishing proteins. Ideal reporter genes should be highly sensitive, stable, and reliable for large-scale measurements within a wide range of cells and tissues. However, the ideal reporter genes encompassing all the desired properties are still unavailable despite current reporter systems being extensively studied. Each reporter system manifests its own beneficial and detrimental traits. Therefore, due consideration would have to be given when contemplating a suitable reporter gene based on the nature of the study.

Problem Posed by Antibiotic Resistance Reporter Genes

Plant transformation techniques available currently are rather efficient but not perfect yet. There are no techniques that are able to provide 100% transformation efficiency. In order to distinguish the transformed and nontransformed plant cells, markers are needed. Antibiotic or herbicide resistance genes act as the primary selective markers in transformant selection to efficiently eliminate the nontransformants. The effectiveness of an antibiotic resistance system is dependent on three criteria:

- Selective agent used should completely inhibit the growth of nontransformed cells.
- Resistance gene is expressed in transformed cells.
- Explant used for transformation.

Table: Selective antibiotics used for transgenic plants screening.

Antibiotics	Mechanism of action	General working concentration (μg/ml)	Selection
Kanamycin	Inhibiting ribosomal translocation and eliciting miscoding	50	nptII
Hygromycin B	Inhibit protein synthesis	20-200	Hph
Streptomycin	Inhibit protein synthesis	100	Spt
Spectinomycin	Inhibit protein synthesis	100	aadA
Phleomycin	DNA breakage	10	ble
Bleomycin	DNA breakage	10	ble

Antibiotic screening has provided the initial identification of successful transgenic plant. However, the use of antibiotics always leads to issues on environmental problems and genetic modified (GM) food safety. This is mainly due to the concern of gene pollution when antibiotic gene escapes from the GM plant into the environment through microorganism. Bacteria are known to be able to uptake and integrate foreign DNA pieces into their genome. The microbes surrounding the GM plant might uptake the DNA fragments from the transgenic plant and hence developed resistance to the antibiotics. Besides, gene escape may occur as the antibiotic resistance gene can be transferred to the neighboring plants through pollen dispersal. There is also the possibility that consuming the transgenic plant with antibiotic resistance gene may result in the transfer of the genes to the pro-bacteria present in the guts. Hence, the antibiotic resistance marker genes are normally avoided in the transgenic whole plant screening.

Efforts had been made by replacing antibiotic marker gene with another potential marker gene such as reporter genes. Reporter genes such as GFP are not reported as toxic to the environment, instead they are widely used as biosensors. Engineering plants with these reporter genes could prevent the unnecessary buildup of antibiotic resistance in the environment. On the other hand, gene escape can also be avoided through the removal of antibiotic resistance gene from the transgenic plant. The latest genome editing tools such as TALEN and CRISPR/Cas9 may be good tools in the removal of gene markers from the transgenic plants. However, these advanced tools have yet to be shaped.

Vectors for the Production of Transgenic Plants

A vector acts as a vehicle that transports the gene of interest into a target cell for replication and expression. Common vector consists of three components: an origin of replication, multicloning site or recombination site, and selectable marker. The origin of replication is an AT-rich region on the vector that initiates the replication of the vector itself by binding to a protein complex, unwinding the vector and thus replicating it with the help of polymerases. The multicloning site is a region that contains multiple unique sequences otherwise known as restriction site that can be cut by specific restriction enzyme, allowing the insertion of the gene of interest. The recombination site allows site-specific recombination to occur between two plasmids. In plant transformation, vectors commonly used are Ti plasmid-based vector and plant viral-based vector.

Plasmid Vectors

Ti Plasmid

The Ti plasmid is the most commonly used vector in the production of a transgenic plant. The Ti plasmid has an estimated size ranging between 200 and 800 kbp depending on the classes of the Ti plasmid. The Ti plasmid is divided into three main regions: the transfer DNA (T-DNA) region, virulence region, and opine catabolism region. The T-DNA region that is transferred into the plant genome is about 24 kbp in size. This region is bordered by repeat sequences on each end commonly known as the left border and right border. The right border is the critical part essential for the transfer of DNA-causing tumorigenesis. The virulence region, however, is responsible for encoding the vir genes, which aids in the transfer of the T-DNA. The T-DNA sequence also codes for opine and phytohormones (auxin and cytokinin) biosynthesis. The three oncogenes (opine, cytokinin, and auxin biosynthesis gene) within the T-DNA are the main causes of tumor formation in plant, leading to the crown gall disease. The growth hormones synthesized are responsible in causing uncontrolled plant cells' proliferation and worsen the situation by enhancing crown gall formation. Opines are the main carbon source utilized by the A. tumefaciens that are not naturally synthesized from plant metabolism. Therefore, A. tumefaciens will develop its own biosynthetic machinery for production of nutrients by genetically modifying the host cells. The opine catabolism region encodes the genes for proteins involved in opines catabolism. The origin of DNA replication allows stable maintenance of the Ti plasmid in the bacterium. For plant transformation, the Ti plasmid is usually disarmed, with the tumor-inducing genes removed and replaced with the reporter genes together with the gene of interest.

The Ti plasmid is large and would become larger with the genes of interest and selectable markers. Large-sized plasmids are cumbersome to handle and have low copy numbers in nature. However, this drawback eventually led to the development of a co-integrative system in combination with the binary vector system which solved the problem for large-sized plasmids.

The Co-integrative Vector

The co-integrative vector is developed through homologous recombination between an intermediate vector and disarmed Ti plasmid. The intermediate vector is normally the E. coli plasmid harboring the gene of interest. Both the intermediate vector and disarmed Ti vector consist of some common sequences, which allow the homologous recombination of the two plasmids to occur. The recombination will result with a large co-integrative vector containing the merged E. coli plasmid and disarmed Ti plasmid. This co-integrative vector will later be introduced back into the Agrobacterium for transgenic plant transformation. However, the enormous size of the plasmid as a result from the recombination may prove an ominous challenge to be manipulated. Thus, the use of this vector had been discontinued since the binary vector system was introduced.

The Binary Vector

A two-plasmid system called the binary vector system was developed when researchers found that T-DNA functioned independently without the needs to attach to the Ti plasmid. The binary system involved two plasmids which are the helper vector and mini vector. The mini vector refers to a smaller size plasmid consisting of the T-DNA and the origin of replication of both E. coli and A. tumefaciens, which allow the plasmid to be cloned in E. coli and A. tumefaciens. The helper vector refers to a wild-type Ti plasmid without the T-DNA region. The wild-type Ti plasmid is also known as a helper plasmid as it provides the template for all the genes necessary for gene transferring and integration. Both of these helper and mini vectors are introduced together into the Agrobacterium and the transformed Agrobacterium will be used in plant transformation.

Plant Virus Vectors

Viruses are intracellular obligate parasites that require molecular machinery from a specific host to replicate. Viruses have not been found to infect plants through the use of transmission vectors such as aphids, insects, nematodes, and fungi. These viruses have been modified and are used as alternative sources for plant transformation; common plant viruses used in transgenic plant production include the Cauliflower mosaic virus (CaMV), Tobacco mosaic virus (TMV), Alfafa mosaic virus (AMV), Potato virus X (PVX), and Cowpea mosaic virus (CPMV). The wild-type plant viral vectors have been improved and modified to accommodate their use with Agrobacteria as well as the plant host for an increased efficiency level through two approaches. The first approach would be designing virus vectors that are similar to wild types carrying the gene of interest, which are capable of infecting plants.

The second approach would be the development of a 'deconstruct' virus, which occurs through the removal of the undesired viral genes, for example, the coat protein-expressing gene, and to replace them with functional gene such as reporter genes or antibiotic resistance gene, which facilitates transgenic screening.

Transformation Techniques

Plant transformation refers to the process of altering the genetic constituents in a plant of interest by introducing DNA segments into the plant genome to achieve desired gene expression. Numerous types of plant transformation techniques have now been made accessible to the public. These plant transformation techniques can be categorized under two groups: indirect or direct gene transfer. Indirect gene transfer (also known as vector-mediated gene transfer) involves the introduction of exogenous DNA into the plant genome via biological vectors, whereas direct gene transfer methods involve the introduction of exogenous DNA directly into plant genome through physical or chemical reactions. Different gene transfer methods and their salient features are tabulated below.

Table: Gene transfer methods in plants and their features.

Method	Features
Vector-mediated gene transfer	
a. Agrobacterium-mediated gene transfer	Efficient to wide range of plants.
b. Plant virus vectors	Efficient and high expression of transgenes.
Direct gene transfer a. Physical Methods	
Electroporation	Confined to protoplasts that can be regenerated to produce complete and viable plants.
ii. Microinjection	Requires highly skillful technical personnel and limited to one cell per microinjection.
iii. Particle bombardment/microprojectile	Special instrumentation required. High risk of gene rearrangement. May be used for a wide range of plant tissues.
iv. Silicon carbide fibers	Requires careful handling. Requires regenerable cell suspensions.
b. Chemical methods	
v. Polyethylene glycol (PEG)-mediated	Confined to protoplasts. Problems encountered when regenerating these cells into viable plants.
vi. Liposome fusion	Confined to protoplasts which may be regenerated into a viable plant.
vii. Diethylaminoethyl (DEAE) dextran mediated	Does not result in stable transformation.

Agrobacterium-mediated Gene Transfer

Agrobacterium-mediated transformation is the most common technique used in plant transformation as it is efficient and effective in a wide range of plants. Agrobacteria are indigenous to the soil ecosystem. They are pathogenic Gram-negative bacteria that cause crown gall or hairy root disease in plants. The genetic information for tumor growth is encoded on a tumor-inducing plasmid (Ti plasmid) or hairy root-inducing plasmid (Ri plasmid) in the genome of these bacteria. There are generally two types of Agrobacterium species that are commonly used in plant transformation; Agrobacterium tumefaciens and Agrobacterium rhizogenes. A. tumefaciens contains the Ti plasmid which causes crown gall disease, whereas A. rhizogenes contains the Ri plasmid that causes hairy root disease. The discovery of these two species provides efficient vector systems for the development of transgenic plants when the detrimental genes in Agrobacteria are removed. This method had successfully transformed a broad variety of plants such as rice, maize, barley, and tobacco.

A. tumefaciens used for plant transformation are modified Agrobacteria which has no tumor-promoting and opine-synthesis genes in their genome. These genes are removed (disarmed) from the bacterial plasmid and replaced with the desired foreign gene or selective markers, making them useful vectors that enables the incorporation of foreign genes into plant's genome, transiently or stably. In order to achieve stable incorporation of genes, the Agrobacteria function to transport and integrate the T-DNA into the host's genome through these steps:

- Chemical signal recognition of host.
- Activation of the vir gene in Agrobacterium.
- Attachment of Agrobacterium to plant cells.
- Activation and transportation of virulence proteins.
- Production of T-DNA strand.
- Transfer of T-DNA and virulence protein out of Agrobacterium.
- Transfer of T-DNA into plant nuclear.
- Integration of T-DNA into plant genome.

The steps involve in transient transformation are postulated to be identical to the stable transformation with the exception of steps.

Ever since the Agrobacterium-mediated transformation protocol had first been introduced, various refinement of the protocol had been ongoing to improve its efficiency. Although traditional Agrobacterium-mediated transformation works efficiently in dicotyledonous plants such as potatoes, tomatoes, and tobacco, it is less successful in recalcitrant crops such as wheat and maize due to the lack of wounding response system. This has remained a critical obstacle until the development of plant tissue culture and the introduction of in-planta transformation protocol that has improved the transformation efficiency on these plants at many folds.

Agrobacterium-mediated in-planta transformation is a method that does not involve plant tissue culture; transformation is done directly onto a developed plant. This technique includes agro inoculation and agro infiltration. In agro inoculation, the transformed Agrobacteria with the gene of interest is inoculated onto the surface of the plant tissue of a whole plant. It is generally done using either by toothpick, wire loop, or direct organ immersion (floral dip method). Agro infiltration, on the other hand, can be carried out using syringe or vacuum. Syringe infiltration is simple and cost effective as it injects the transformed Agrobacteria onto the underside of the leaf while concurrently ensuring the application of counter pressure on the other side of the leaf. However, it is time consuming and only suitable for small-scale expression.

Vacuum infiltration, however, is rapid and more efficient, thus enabling large-scale production intended commercialization. In this approach, the whole plant is submerged into the transformed Agrobacteria suspension with application of a vacuum environment that forces the Agrobacteria to penetrate throughout the whole plant. These alternative methods have become popular in plant transformation especially in monocots.

Heat and hydrolysis treatment on target tissues prior to transformation have been reported to enhance transformation efficiency when heat treatment is used, enhancing the efficiency of transformation in different plant tissue such as switchgrass, ryegrass, and rice. Similar enhancement had also been obtained in hydrolysis treatment via hydrolytic enzyme such as cellulase, macerase, and pectinase, which provides milder disruption that improved the recovery and regeneration rate of transformed cells. In addition, sonication-assisted plant transformation applied to the target plant tissues prior to Agrobacteria immersion or agroinfiltration resulted in effective transformation of recalcitrant plants by creating micro-wound on explants, which provides better access to Agrobacterium.

Direct Gene Transfers

Direct gene transfer, as the name suggests, involves the direct introduction of exogenous DNA (naked DNA) into the plant nucleus. In order to introduce foreign DNA into the plant cell, the outer membrane of the cell is first disrupted, permeating it for foreign DNA to enter. Most of the methods under direct gene transfer are simple and effective. However, gene expression in these transgenic plants can be transiently or stably transformed.

Direct gene transfer can be categorized into two main groups: physical gene transfer and chemical gene transfer. Physical gene transfer disrupts the cell wall and cell membrane via mechanical means. Among these methods, particle bombardment biolistic is the most common one used in plant transformation since it was first introduced by Sanford et al. The DNA coated with gold or tungsten particles are shot into the target plant cell under high pressure using a "Gene Gun" (Helios® Bio-Rad). The fast-moving particles allow for the penetration of coated DNA through the thick plant cell wall, directing the foreign DNA into its nucleus. The coated DNA will then separate from the metal particles and integrate itself into the chromosomes within the nucleus of the plant cell. This method had been found to be effective in transforming both dicots and monocots which compensates for the less successful Agrobacterium-mediated transformation process. Furthermore, it is also less toxic and applicable to almost all plant cells. The major setbacks of this method, however, lie in the availability of special instruments as well as the delivery efficiency of DNA fragments to the plant nucleus instead of other organelles. In traditional biolistic method, microprojectiles (gold or tungsten) are normally coated with DNA in the presence of calcium chloride and spermidine. The spermidine helps to stabilize the DNA structure and enhances the binding of DNA to the microprojectiles. In the effort of improving this tool, other cationic polyamine such as protamine provides better results when compared with spermidine, as this ensures by protecting the coated DNA from DNase degradation. Biolistic transformation via protamine had been performed in rice and peanut, and the results were shown to be threefolds better when compared to spermidine. Other methods of improving the efficiency of transformation via biolistic guns involved reduction of the amount of DNA coated on the micro carriers.

Other physical gene transfer methods include electroporation that uses electrical impulses to facilitate the transfer of foreign DNA into the plant cells. Plant cells are first incubated in a buffer solution containing foreign DNA, followed by the application of electrical impulses into the buffer, resulting in the formation of temporary transient pores on the cell membrane of the plant to allow the foreign DNA to enter. This method is relatively easy and time saving but is only applicable to protoplasts (cell without cell wall). Hence, this method is not commonly practiced in plant transformation.

Chemical gene transfer approaches involves the use of chemical to disrupt cell membrane enabling the entry of foreign DNA. This particular method is not preferable in plant transformation as it is only effective when applied to protoplasts. One of the most prominent chemicals used in this approach is polyethylene glycol (PEG) that is used for destabilizing the cell membrane in the presence of a divalent cation, thus increasing the permeability of the cell membrane, allowing for the uptake of foreign DNA. The exact mechanism for chemical gene transfer is not fully understood, but it was postulated that PEG increases the osmotic pressure and causes contraction in the protoplast; this facilitates endocytosis of the divalent cation/DNA complex. Besides those, liposome is another chemical method that is used in the transformation of plant's protoplast cells. Liposomes act as vehicles to encapsulate and deliver foreign genetic materials into the protoplast. The lipophilic attribute of liposomes provide easy access into the protoplast in transforming the cell.

Integration and Inheritance of the Transgenes

Integration of Transgenes

Integration of transgenes into plant cells can be carried out either stably or transiently. In stable transformation, the process normally begins with the introduction of transgenes into the nucleus of plant cells. Stable transformation is achieved when some of the transgenes integrate successfully into the genome of the cell. These transgenes then become a part of the genome and are replicated together, enabling the next generation to inherit and express the transgene. In contrast, transient transformant expressed the transgene transiently, and the transgene is not integrated into the plant genome. In the transiently transformed plant, the copy numbers of transgene inserted remain as they are not replicated. These transgenes are expressed for a limited period of time, and the genes will be lost after several days through cell division. The way how the transgene is expressed in the cell is dependent on the transgene construct design and the method of transformation used. Currently, transient and stable transformation can be achieved through the Agrobacterium-mediated method and biolistic method. In the Agrobacterium-mediated method, the T-DNA region is inserted into the plant genome forming a stable transformant, whereas the non-integrated T-DNA plasmid expresses the transgene transiently. In the biolistic method or other direct gene transfer methods such as electroporation, transient and stable expression of the transgenes are usually dependent on the plasmid or transgene constructs. Virus-mediated vectors are generally non-integrative vectors for which transient transformants are frequently produced.

Inheritance of Transgenes

Inheritance of plant genetic information usually obeys the Mendelian law of inheritance in nature.

Mendel's first law, the principle of segregation, states that a pair of alleles for each gene will segregate during the formation of gametes, resulting in each gamete harboring only one allele of the gene. Mendel also discovered that the genes of different traits assort independently of each other in the formation of gametes; these genes are passed down to the subsequent progeny generation according to the rules of probability. In addition, the third Mendelian's law states that one allele is dominant to the other allele, which finally determines the corresponding phenotypic attribute of the offspring. However, there are certain cases in which inheritance of a gene does not comply with Mendel's law (non-Mendelian law). These instances include incomplete dominance, co-dominance, gene controls by multiple alleles, and polytraits.

Similarly, transgene inheritance may or may not obey the Mendelian law. The rule for transgene inheritance, however, varies due to the location of transgene integration and the copy number of transgenes integrated. Transgene inheritance not obeying the Mendelian law includes deletion of the transgene locus, rearrangement of the inserted transgene, and silencing of the transgene. Overall, the pattern of transgene inheritance is usually analyzed through molecular characterization of the transgene transmission and the segregation analysis of the transgene phenotypic expression pattern.

Table: Factors leading to non-Mendelian inheritance of transgene.

	Factors
Nature of recipient genome	Genetic background Gamete viability Chromosome abnormality Transformation method
Nature of transgene	Transgene silencing Unstable integration of transgene
Interactions between the recipient genome and the transgene	Homozygous lethality Poor transmission of transgene Mitotic crossover Meiotic instability

Analysis and Confirmation of Transgene Integration

Analysis and confirmation of transgene integration has to be done through an appropriate method based on the transgene constructs, selectable marker, and reporter gene used. Transgenic plant cells incorporated with ant herbicidal or antibiotic resistance genes are screened by the addition of herbicides or antibiotics to the growing media to distinguish transformed plant cells from the non-transformed plant cells. However, this method requires a large quantity of antibiotics and herbicides that are expensive and worsen by the risk of horizontal gene transfer to other bacteria. Thus, other screening methods such as polymerase chain reaction (PCR) and reporter gene expression screening are used for better accuracy as an alternative screening method for transgenic plants.

Some reporter genes such as the GFP, GUS, and Luc expression are fluorometric or colorimetric, where the expression of these genes could be observed visually or directly under microscopy.

Quantifications of the reporter expression are possible with the use of a spectrophotometer. The GUS expression can also be detected through histochemical assay in which the localization of the transgene can be observed. In addition, some of the reporter gene expressions such as CAT and LacZ activity are screened through enzyme assays.

Southern blotting is a molecular method used for the detection of specific DNA sequences within DNA samples. Southern blotting is generally used to identify the number of transgenes inserted into the host genome as well as for the detection of transgene integrity and transgene rearrangement. It is done by cutting the DNA into fragments with endonuclease restriction enzymes, separation by size through electrophoresis, and subsequently transferred onto a nitrocellulose or nylon membrane. Membranes with bound DNA will be incubated in a solution consisting labeled probes, and the pattern of hybridization is detected through autoradiography or via chromogenic detection. The transgene copy number is proportional to the number of bands observed.

The polymerase chain reaction (PCR) method is one of the most sensitive and easiest methods among all the molecular techniques employed for the verification of the transgene. The PCR is generally done with primers specific to the site of plasmid constructs and gene of interest used for development of the transgenic plants. Successful amplification of the DNA fragment with expected band indicates the possible presence of transgene, and this DNA fragment is further confirmed through DNA sequencing. A real-time PCR provides fast, sensitive, and high-throughput molecular PCR–based analysis compared to the traditional Southern blot analysis especially in the area of transgene copy number and zygosity detection in transgenic plants. Real-time PCR is convenient wherein it allows for quantitative, semi-quantitative (qPCR), or qualitative (RT-qPCR) monitoring of target DNA in real time.

In recent years, the emergence of next-generation sequencing (NGS) technologies allows massive parallel generation of sequences from whole genome in a relatively short time with a lower cost. The PCR-based techniques in transgenic analysis often limits by the generation of non-specific products and failure to amplify large exogenous DNA insertion in highly repetitive genomes, multiple insertion, truncated transgene sequences and hinders precise transgene identification. The availability of NGS tools and bioinformatics resources facilitate the study of genome and molecular characterization of complex traits. Besides, the analyses of NGS data allow the identification of precise genomic locations of transgene insertion especially in highly repetitive genome sequence and transposable elements which was not able to be done through the traditional PCR-based method. Hence, NGS approach provides an alternative high-resolution analysis tool for transgenes insertion in GM crops.

Haploids and Doubled Haploids in Plant Breeding

Haploids are plants (sporophytes) that contain a gametic chromosome number (n). They can originate spontaneously in nature or as a result of various induction techniques. Spontaneous development of haploid plants has been known since 1922, when Blakeslee first described this phenomenon in Datura stramonium; this was subsequently followed by similar reports in tobacco (Nicotiana tabacum), wheat (Triticum aestivum) and several other species. However, spontaneous occurrence is a rare event and therefore of limited practical value. The potential of haploidy for plant breeding arose in 1964 with the achievement of haploid embryo formation from in vitro culture of Datura anther, which was followed by successful in vitro haploid production in tobacco. Many attempts have been made since then, resulting in published protocols for over 250 plant species belonging to almost all families of the plant kingdom. In fact, under optimal conditions, doubled haploids (DH) have been routinely used in breeding for several decades, although their common use is still limited to selected species. There are several reasons for this. These might be categorized as biological, based on plant status (annual, biannual, perennial, authogamous, allogamous, vegetativelly propagated) and flower morphology or technical, which are the result of the feasibility and efficiency of DH induction protocol. Induction protocols substantially vary, in fact, not only among species but also among genotypes of the same species.

Production of Haploids and Doubled Haploids

Haploids produced from diploid species (2n=2x), known as monoploids, contain only one set of chromosomes in the sporophytic phase (2n=x). They are smaller and exhibit a lower plant vigor compared to donor plants and are sterile due to the inability of their chromosomes to pair during meiosis. In order to propagate them through seed and to include them in breeding programs, their fertility has to be restored with spontaneous or induced chromosome doubling. The obtained DHs are homozygous at all loci and can represent a new variety (self-pollinated crops) or parental inbred line for the production of hybrid varieties (cross-pollinated crops). In fact, cross pollinated species often express a high degree of inbreeding depression. For these species, the induction process per se can serve not only as a fast method for the production of homozygous lines but also as a selection tool the elimination of genotypes expressing strong inbreeding depression. Selection can be expected for traits caused by recessive deleterious genes that are associated with vegetative growth. Traits associated with flower fertility might not be related and should be eliminated by recurrent selection among DH lines.

The production of pure lines using doubled haploids has several advantages over conventional methods. Using DH production systems, homozygosity is achieved in one generation, eliminating the need for several generations of self-pollination. The time saving is substantial, particularly in biennial crops and in crops with a long juvenile period. For self-incompatible species, dioecious species and species that suffer from inbreeding depression due to self-pollination, haploidy may be the only way to develop inbred lines.

The induction of DH lines in dioecious plants, in which sex is determined by a regulating gene, has an additional advantage. Such a case is well studied in asparagus; in which sex dimorphism is determined by a dominant gene M. Female plants are homozygous for the recessive alleles (mm), while male plants are heterozygous (Mm). Androgenically produced DH lines are therefore female (mm) or 'supermale' (MM). An advantage of supermales is that, when used as the pollinating line, all hybrid progeny are male.

Haploids from polyploid species have more than one set of chromosomes and are polyhaploids; for example dihaploids (2n=2x) from tetraploid potato (Solanum tuberosum ssp. tuberosum, 2n=4x), trihaploids (2n=3x) from heksaploid kiwifruit (Actinidia deliciosa, 2n=6x) etc. Dihaploids and trihaploids are not homozygous like doubled haploids, because they contain more than one set of chromosomes. They cannot be used as true-breeding lines but they enable the breeding of polyploid species at the diploid level and crossings with related cultivated or wild diploid species carrying genes of interest.

The main factors affecting haploid induction and subsequent regeneration of embryos are:

- Genotype of the donor plants.
- Physiological condition of donor plants (i. e. growth at lower temperature and high illumination).
- Developmental stage of gametes, microspores and ovules.
- Pre-treatment (i. e. cold treatment of inflorescences prior to culture, hot treatment of cultured microspores).
- Composition of the culture medium (including culture on "starvation" medium low with carbohydrates and/or macro elements followed by transfer to normal regeneration medium specific to the species).
- Physical factors during tissue culture (light, temperature).

Haploid Techniques

In Situ Induction of Maternal Haploids

In situ induction of maternal haploids can be initiated by pollination with pollen of the same species (e.g., maize), pollination with irradiated pollen, pollination with pollen of a wild relative (e.g., barley, potato) or unrelated species (e.g., wheat). Pollination can be followed by fertilization of the egg cell and development of a hybrid embryo, in which paternal chromosome elimination occurs in early embryogenesis or fertilization of the egg cell does not occur, and the development of the haploid embryo is triggered by pollination of polar nuclei and the development of endosperm.

Pollination with Pollen of the Same Species

Maternal haploid induction in maize (Zea mays L.) is a result of legitimate crossing within one species with selected inducing genotypes (line, single cross or population). It results in a majority of regular hybrid embryos and a smaller proportion of haploid maternal embryos with normal triploid endosperms. The first recognized inducer line was the genetic strain Stock 6, with an haploid induction rate of up to 2.3%, which was subsequently improved by hybridization and further selection. Today, modern haploid inducing lines display high induction rates of 8 to 10%. They are routinely used in commercial DH-line breeding programs due to their high effectiveness and lower genotype dependence. In contrast to other induction techniques, no in vitro culture is needed, since kernels containing haploid embryos display a normal germination rate and lead to viable haploid seedlings. Haploid embryos can be selected early in the breeding process, based on morphological and physiological markers.

Pollination with irradiated pollen is another possibility for inducing the formation of maternal haploids using intra-specific pollination. Embryo development is stimulated by pollen germination on the stigma and growth of the pollen tube within the style, although irradiated pollen is unable to fertilize the egg cell. It has been used successfully in several species.

The production of maternal haploids stimulated by irradiated pollen requires efficient emasculation, which has in some cases been shown to limit its use because the method is too laborious. To overcome such an obstacle in onion, for instance, only male sterile donor plants were used as donor plants, but such lines, possessing cytoplasmically inherited male sterility, are of very limited practical use. Apart from the factors affecting haploid production already mentioned, the dose of irradiation is the main factor controlling in situ haploid production. At lower doses, the generative nucleus is partly damaged and therefore maintains its capacity to fertilize the egg cell. It results in large numbers of obtained embryos but all of hybrid origin and abnormal (mutant) phenotype. An increase in the irradiation dose causes a decrease in the total number of developed embryos but the obtained regenerants are mostly of haploid origin.

For most plant species, in vitro embryo rescue is necessary to recover haploid plants. The collection of mature seeds has only been reported for kiwifruit, onion, mandarin and species of the genus Nicotiana. Even for the aforementioned species, in vitro germination of seeds enhanced the recovery of haploid plants.

Wide Hybridization

Wide crossing between species has been shown to be a very effective method for haploid induction and has been used successfully in several cultivated species. It exploits haploidy from the female gametic line and involves both inter-specific and inter-generic pollinations. The fertilization of polar nuclei and production of functional endosperm can trigger the parthenogenetic development of haploid embryos, which mature normally and are propagated through seeds (e.g., potato). In other cases, fertilization of ovules is followed by paternal chromosome elimination in hybrid embryos. The endosperms are absent or poorly developed, so embryo rescue and further in vitro culture of embryos are needed (e.g., barley).

In barley, haploid production is the result of wide hybridization between cultivated barley (Hordeum vulgare, 2n=2x=14) as the female and wild H. bulbosum (2n=2x=14) as the male. After fertilization, a hybrid embryo containing the chromosomes of both parents is produced. During early embryogenesis, chromosomes of the wild relative are preferentially eliminated from the cells of developing embryo, leading to the formation of a haploid embryo, which is due to the failure of endosperm development. A haploid embryo is later extracted and grown in vitro. The 'bulbosum' method was the first haploid induction method to produce large numbers of haploids across most genotypes and quickly entered into breeding programs. Pollination with maize pollen could also be used for the production of haploid barley plants, but at lower frequencies.

Paternal chromosome elimination has also been observed after interspecific crosses between wheat (Triticum aestivum) and maize. After pollination, a hybrid embryo between wheat and maize develops but, in the further process, the maize chromosomes are eliminated so that haploid wheat plantlets can be obtained. Such haploid wheat embryos usually cannot develop further when left on the plant, because the endosperm fails to develop in such seeds. By applying growth regulator 2, 4-dichlorophenoxyacetic acid in planta, embryo growth is maintained to the stage suitable for embryo isolation and further in vitro culture. The maize chromosome elimination system in wheat enables the production of large numbers of haploids from any genotype. Pollination with maize is also effective for inducing haploid embryos in several other cereals, such as barley, triticale (x Triticosecale), rye (Secale cereale) and oats (Avena sativa). Similar processes of paternal chromosome elimination occur after the pollination of wheat with wild barley (H. bulbosum), sorghum (Sorghum bicolour L. Moench) and pearl millet. In contrast to maize and pearl millet pollination, pollination with H. bulbosum is strongly influenced by the maternal genotype.

Haploid production in cultivated potato (Solanum tuberosum L. ssp. tuberosum, 2n=4x) can be achieved by inter-generic pollination with selected haploid inducer clones of S. phureja (2n=2x). The tetraploid female S. tuberosum produces an embryo sac containing one egg cell and two endosperm nuclei, all with the genetic constitution n=2x, while the diploid pollinator S. phurea produces two sperms of the genetic constitution n=x or 2x. After pollination, dihaploid (2n=2x) embryos can develop from un-fertilized egg cells, which are supported by a 6x endosperm formed by the fusion of polar nuclei with both reduced sperm cells. The frequency of dihaploid seeds is low; they have to be selected from hybrid seeds containing 3x or 4x embryos developed from egg cells (n=2x) fertilized with haploid (n=x) or diploid (n=2x) sperm cells. Dihaploid potatoes can be used for breeding purposes, including alien germplasm introgression or selection at the diploid level, but such plants are not homozygous. Haploids have a significant role in potato breeding programs of quite a few companies, since they enable interspecific hybridization, which would not otherwise be possible due to differences in ploidy levels and endosperm balance numbers. The gene pool of potato can be broadened and certain valuable traits, such as disease resistance characters from the wild solanaceous species, can be more efficiently introgressed into cultivated potato.

In Vitro Induction of Maternal Haploids – gynogenesis

In vitro induction of maternal haploids, so-called gynogenesis, is another pathway to the production of haploid embryos exclusively from a female gametophyte. It can be achieved with the in vitro culture of various un-pollinated flower parts, such as ovules, placenta attached ovules, ovaries or whole flower buds. Although gynogenetic regenerants show higher genetic stability and a lower rate of albino plants compared to androgenetic ones, gynogenesis is used mainly in plants in which other induction techniques, such as androgenesis and the pollination methods above described, have failed. Gynogenic induction using un-pollinated flower parts has been successful in several species, such as onion, sugar beet, cucumber, squash, gerbera, sunflower, wheat, barley etc., but its application in breeding is mainly restricted to onion and sugar beet.

The success of the method and its efficiency is greatly influenced by several biotic and abiotic factors. The genotype of donor plants, combined with growth conditions, is the crucial factor. In onion, for example, pronounced differences in embryo yields have been recorded among accessions and among plants within accessions. The average frequencies of induced embryos (calculated from ovaries possessing 6 ovules) varied between 0% in non-responding accessions to 18.6-22.6% in extremely responsive accessions, with individual donor plants producing up to 51.7% embryos. The high haploid production frequency was tested in two consecutive years and showed to be stable over years. Induction rates were even higher in preselected onion genotypes, achieving frequencies of 196.5%embryos from a doubled haploid line or 82.2% for an inbred line.

Production of onion haploid plants with in vitro gynogenesis. (A) In vitro culture of un-pollinated flower buds on BDS medium supplemented with 500 mg/l myo-inositol, 200 mg/l proline, 2 mg/l BAP, 2 mg/l 2,4-D, 100 g/l sucrose and 7 g/l agar; (B) germination of haploid embryos after 60 to 180 days in culture; (C) elongation of haploid plantlets and (D) acclimatization of haploid plants in the greenhouse.

Developmental stage of gametes, the pre-treatment of flower buds prior to inoculation, in vitro culture media and culture conditions are other factors affecting the embryogenic response of gametes in culture. The female gametophyte is usually immature at inoculation and, in contrast to androgenesis, its development continues during in vitro culture, leading to a mature embryo sac. Mature embryo sacs contain several haploid cells theoretically capable of forming haploid embryos, such as the egg cell, synergids, antipodal cells and non-fused polar nuclei. However, under optimal conditions, the egg cells in most gynogenetic responsive species undergo sporophytic development (haploid parthenogenesis). They can develop into haploid plants directly, avoiding the risk of gametoclonal variation, or through an intermediate callus phase.

Media components, mainly the type and concentration of carbohydrates and plant growth regulators, play an important role in reprogramming haploid cells from gametophytic to the sporophytic pathway. The requirements are species and genotype dependent and no universal protocol for in vitro gynogenesis exists.

Induction of Paternal Haploids – Androgenesis

Androgenesis is the process of induction and regeneration of haploids and double haploids originating from male gametic cells. Due to its high effectiveness and applicability in numerous plant species, it has outstanding potential for plant breeding and commercial exploitation of DH. It is well established for plant breeding, genetic studies and/or induced mutations of many plant species, including barley, wheat, maize, rice, triticale, rye, tobacco, rapeseed, other plants from Brassica and other genera. Its major drawbacks are high genotype dependency within species and the recalcitrance of some important agricultural species, such as woody plants, leguminous plants and the model plant Arabidopsis thaliana. The method relies on the ability of microspores and immature pollen grains to convert their developmental pathway from gametophytic (leading to mature pollen grain) to sporophytic, resulting in cell division at a haploid level followed by formation of calluses or embryos.

Androgenesis can be induced with in vitro culture of immature anthers, a technically simple method consisting of surface sterilization of pre-treated flower buds and subsequent excision of anthers under aseptic conditions. The anthers are inoculated and cultured in vitro on solid, semi-solid or liquid mediums or two-phase systems (liquid medium overlaying an agar-solidified medium). Anther culture was the first discovered haploid inducing technique of which efficiency was sufficient for plant breeding purposes. It is still widely used, although isolated microspore culture is an improved alternative. During isolation of microspores, the anther wall tissues are removed, thus preventing interference of maternal sporophytic tissue during pollen embryogenesis and regeneration from somatic tissue. Moreover, basic research of haploid embryogenesis can be performed directly at the cellular, physiological, biochemical and molecular levels.

Androgenesis, like other haploid inducing techniques, is influenced by several biotic and abiotic factors. The developmental stage of male gametes at the time of anther or microspore isolation, in combination with suitable stress treatments, are the main factors determining the androgenetic response. It can be triggered within a relatively wide developmental window around the first pollen mitosis, when uninucleate microspores divide asymmetrically resulting in a generative cell embedded in a vegetive cytoplasm.

The application of suitable physiochemical factors promotes a stress response, which arrests the microspores or young pollen grains in their gametophytic pathway. Their development is triggered through embryogenesis by promoting cell divisions and the formation of multicellular structures contained by the exine wall. Finally, the embryo-like structures are released from the exine wall. The most widely used triggering factors are temperature pre-treatment, sucrose and nitrogen starvation and osmotic stress. Depending on the plant species and genotype, temperature stress can be applied by subjecting excised flower buds, whole inflorescences or excised anthers to low (barley, wheat, maize, rice, triticale, rye) or high (rapeseed, Brassica species, tobacco, wheat) temperatures for several hours or days. As demonstrated in rapeseed and tobacco, different triggering factors can promote embryogenesis from microspores or immature pollen cells at different developmental stages. In rapeseed, early binucleate pollen grains can be converted to the embryologic pathway by applying a heat shock treatment at 32 °C, while late binucleate pollen needs an extra heat shock treatment at 41 °C. In tobacco, a heat shock treatment is effective in triggering unicellular microspores but not in triggering immature bicellular pollen grains, which successfully start embryogenesis after sucrose and nitrogen starvation.

Figure: Microspore culture of cabbage: (A) first divisions of microspores in NLN medium, (B) regenerated embryos, (C) embryos at desiccation treatment needed for regrowth, (D) selfing of DH lines.

Several other triggering factors such as irradiation, colchicine, auxin and water stress are also used for reprogramming microspores, but to a limited extent. The androgenetic response can be enhanced by in vivo pre-treatments of donor plants with nitrogen starvation, short days and low temperature culture conditions.

In addition to stress treatments, the majority of studies have focused on culture media constituents. In general, the concentration of salts in the culture media is lower compared to micro propagation media, but there is no general rule. Several commonly used recipes of macro and micro elements are often used, such as potato-2, W-14, NLN, A2 medium and others. The same media are often used for systematically diverse taxa, for instance NLN medium developed for Brassica napus was also efficient in Apiaceae , while A2 medium developed for tobacco was also optimal for wheat. Occasionally, even an increased concentration of micronutrients might be of high value, for instance elevated copper concentration reduced albinism in cereals. Choices and concentrations of carbohydrates are often essential. The most commonly used carbohydrate is sucrose, particularly in microspore media, and is added in high concentrations (i.e., 13%), while substitution of sucrose by maltose has been an important innovation, first discovered for barley anther culture. Although not always required, plant growth regulators might be essential.

Under optimal in vitro culturing conditions, androgenetic plants are regenerated from embryo-like structures (direct microspore embryogenesis) or from microsporial callus cells (organogenesis). Direct embryogenesis is preferred, since regeneration through the callus stage might induce undesired gametoclonal variation and might also cause albinism.

Identification of (Doubled) Haploids: Ploidy Level Determination and Homozygosity Testing

After a successful haploid induction and the regeneration procedure, evaluation of regenerants is needed to distinguish between desired haploids (or spontaneously doubled haploids) and redundant heterozygous diploids. The haploid inducing techniques presented here differ in haploid induction rates and in the type of undesired regenerants which can be obtained. In this regard, isolated microspore culture is superior to other techniques because the filtering of microspore suspensions during isolation for the most part prevents diploid plant residuals from entering into the in vitro culture and later the regeneration of heterozygotes. Moreover, a relatively high proportion of induced haploids spontaneously double their chromosome number, leading to regeneration of homozygous doubled haploids. The phenomenon of spontaneous diploidization of microspore derived embryos has been studied in more detail in barley. In microspores pretreated with mannitol, chromosome doubling is caused by nuclear fusion after the first nuclear division. In contrast to microspore culture and androgenesis in general, gynogenic regeneration leads to predominantly haploid regenerants in the majority of species.

During the production of homozygous lines, various undesired heterozygous plantlets can be obtained. In anther culture and in vitro gynogenesis, such plants can be regenerated from the somatic tissue of inoculated plant organs such as anther wall cells, somatic cells of flower buds, ovaries or ovules. Moreover, heterozygous hybrids, produced after wide hybridization or after pollination with irradiated pollen, are another class of regenerants that could negatively affect breeding progress, if not discarded. Reliable and fast selection of regenerants is therefore necessary before further employment of putative haploids and doubled haploids.

Several direct and indirect approaches are available for determining the ploidy level of regenerated plants. Indirect approaches are based on comparisons between regenerated and donor plants in terms of plant morphology (plant height, leaf dimensions and flower morphology), plant vigor and fertility, number of chloroplasts and their size in stomatal guard cells. They are fairly unreliable and subject to environmental effects but do not require costly equipment. Direct methods for ploidy determination are more robust and reliable and include conventional cytological techniques, such as counting the chromosome number in root tip cells and measurement of DNA content using flow cytometry. The latter provides a rapid and simple option for large-scale ploidy determination as early as in the in vitro culturing phase. It also enables detection of mixoploid regenerants (having cells with different ploidy) and the determination of their proportion.

Regeneration of diploid plants is not always caused by unwanted adventitious regeneration from somatic cells or germination of hybrid embryos. Spontaneously doubled haploids may also occur, thus eliminating the need for chromosome doubling. Several markers can be used for assessing the origin of diploids, depending on their availability for a particular plant species. In the past, evaluation of regenerants mainly relied on phenotypic markers, progeny testing after self-pollination and isozyme analysis. Nowadays, DNA molecular markers, such as AFLP (Amplified Fragment Length Polymorphism), RAPD (Random Amplified Polymorphic DNA), SCAR (Sequence Characterized Amplified Regions) or SST (Simple Sequence Repeat), are commonly used for homozygosity testing and assessment of plant origin. There is a considerable difference in interpretation between dominant or co-dominant electrophoretic profiles. Co-dominant molecular markers, as well as isozyme markers, have the advantage that a single locus, when heterozygous in donor plants, might be used for homozygosity determination. In contrast, a more complex profile is analyzed with dominant markers. In such a case, bands missing from the donor profile indirectly indicate homozygosity.

A fast and reliable haploid identification method is needed for large scale production of DHs. Morphological markers expressed at the embryo, seed or early seedling stages are preferentially used. In maize, the most efficient haploid identification marker is the 'red crown' kernel trait, which causes deep pigmentation of the aleurone layer in the crown region (endosperm) and scutellum (embryo tissue). In a haploid inducing cross, the marker should be homozygous recessive in the female parent and homozygous dominant in the pollinator inducer line. After pollination, kernels with a red aleurone crown (resulting from regular fertilization of polar nuclei) containing a nonpigmented scutellum are visually selected from the hybrid kernel of regular fertilization with both aleurone and scutellum pigmented. A similar approach is used in potato, in which selection is based on a homozygous dominant color marker gene carried by the pollinator line. The purple spot embryo marker shows up on seeds whose embryos possess a genome from the pollinator. Those hybrid seeds are discarded, while spotless dihaploid seeds are included in breeding process. Selection can be repeated at the seedling stage, when a purple nodal band can be detected on the hybrid's stem. In the case of both maize and potato selectable markers, it is not possible to distinguish hybrid seeds resulting from unintentional self-pollination of donor plants. Selection has to be supplemented with other morphological or molecular markers.

Chromosome Doubling

Following regeneration, haploid plants obtained from either anther or ovule culture may grow normally under in vitro conditions or can even be acclimatized to form vital mature plants. Such plants often express reduced vigor but in some crops such as onion, even haploid plants might grow vigorously. At the flowering stage, haploid plants form inflorescences with evident malformations. Due to the absence of one set of homologous chromosomes, meiosis cannot occur, so there is no seed set. Duplication of the chromosome complement is therefore necessary.

In pollen derived plants, spontaneous duplication of chromosomes may occur in cultures, often in a sufficient proportion, thus eliminating the need for chemically induced doubling. Spontaneously doubled plants are sometimes preferred because of the fear that the duplication process might induce undesired mutations. Mechanisms of spontaneous doubling differ, with nuclear fusion being the most common cause. As first described by Sunderland et al., synchronous division of two or more nuclei in early stages of embryo development might develop a common spindle. The nuclear fusion theory is supported by the frequent occurrence of a small proportion of triploid regenerants. Nuclear fusions might be associated with delayed cell wall formation, which, as reviewed by Kasha, is typical of cereals. Other mechanisms such as endomitosis are another possible mechanism that is currently less understood. For maternally derived haploid plants, the rate of spontaneous doubling is often much lower or entirely absent, so a doubling procedure is essential.

Various methods have been applied over several decades and are still in development. The most frequently used application is treatment with anti-microtubule drugs, such as colchicine (originally extracted from autumn crocus Colchicum autumnale), which inhibits microtubule polymerization by binding to tubulin. Although colchicine is highly toxic, used at a millimolar concentration and known to be more efficient in animal than in plant tissues, it is still the most widely used doubling agent. Other options are oryzalin, amiprophosmethyl (APM), trifluralin and pronamide, all of which are used as herbicides and are effective in micromolar concentrations. Anti-microtubule drugs might be applied at various stages of androgenesis, such as being incorporated into microspore pretreatment media. Colchicine application on anther culture medium, for instance, showed a significant increase in embryo formation and green plant regeneration in wheat . More often, duplication treatments are applied after regeneration at either embryo, shoot or plantlet level. Similarly, treatments of gynogenically derived embryos with colchine have also been found to be appropriate. The treatment of plants at later developmental stages has the advantage that only already tested haploid regenerants are treated either in vitro (for instance at the shoot culture stage) or in vivo following acclimatization.

The concentration and duration of treatments must be always determined in relation to two effects: the percentage of doubled plants and the percentage of survival. Optimization treatments often require large experimental units (such as 300 explants per treatment) due to the substantial variation of response. High doses/durations can lead to tetraploidization.

Treatment with nitrogen oxide (N_2O), which was developed for maize seedlings is a special case. Plantlets are treated at a high pressure of 600 kPa for two days at the six-leaf stage, in which plants develop flower primordia. The mechanism of action was studied by Kitamura et al. in Lillium and depolymerization of microtubules was found to be the cause.

Chemical treatment might be avoided by using in vitro adventitious somatic regeneration, which itself frequently leads to increased ploidy. Such an approach was efficient in onion. The method has two advantages: the first being that no potentially damaging chemicals are used in the process and the second that regenerants do not for the most part show a mixoploid character. Up to 100% doubling efficiency in relation to individual line treatment can be achieved using this method.

Applications of Doubled Haploids in Plant Breeding

The induction and regeneration of haploids followed by spontaneous or induced doubling of chromosomes are widely used techniques in advanced breeding programs of several agricultural species. They have been successfully used for commercial cultivar production of species such as asparagus, barley, Brassica juncea, eggplant, melon, pepper, rapeseed, rice, tobacco, triticale, wheat and more than 290 varieties have already been released. Using DH technology, completely homozygous plants can be established in one generation thus saving several generations of selfing comparing to conventional methods, by which also only partial homozygosity is obtained.

Another feature that should be considered is the breeding strategy. Within the breeding process, DH lines can be induced as soon as from F1 generation, although some breeders prefer to induce DH lines from later generations. Induction in the F2 generation was proposed as an option because lines originated from F3 generation gametes had passed through another recombination cycle. However, Choo et al., comparing DH and single seed descent methods showed that there was no difference in the sample of recombinants.

The role of DH in the breeding process largely depends on the plant mode of reproduction. In self-pollinated species, they can represent final cultivars or they can be used as parental lines in hybrid production or test-crosses of cross-pollinated species. The basic breeding scheme in self-pollinated species starts with crossings of desired genotypes, leading to hybrids containing chromosome sets of both parents. During gamete formation, recombination enable new gene combinations, which are fixed in the process of doubled haploid production. Doubled haploids thus represent recombinant products of parental genomes in a completely homozygous state. They can be propagated as true breeding lines, facilitating large-scale testing of agronomic performance over the years. Due to complete homozygosity, the efficiency of selection for both qualitative and quantitative characters is increased since recessive alleles are fixed in one generation and directly expressed. Additionally, doubled haploids can be used in a recurrent selection scheme in which superior doubled haploids of one cycle represent parents for hybridization for the next cycle. Several cycles of crossing, doubled haploid production and selection are performed and gradual improvement of lines is expected due to the alternation of recombination and selection.

Similarly as with self-pollinated species, the use of doubled haploids in cross-pollinated species improves selection efficiency and can be used at any or each cycle of recurrent selection. Cross-pollinated species are known to possess numerous deleterious recessive alleles that are not expressed in heterozygous states. They are gradually fixed during selfpollination, causing inbreeding depression and difficulties in producing homozygous lines during conventional breeding. Doubled haploid technology helps to overcome these problems through the rapid fixation of genes in one generation and early elimination of deleterious alleles from populations. The recovered recombination products thus represent more viable combinations of genes. Their complete homozygosity enables true breeding and stable field performance over generations of progeny, although the complete lack of heterozygosity and heterogeneity in varieties is thought to be more vulnerable to environmental changes and altered cropping systems.

It should be noted that, following chromosome doubling, DH plants are normally selfed for maintenance and for further multiplication. In cross-pollinated species with strongly expressed self-incompatibility, various techniques are used to overcome the incompatibility reaction. For instance in Brassicas, bud pollination is enhanced by treatment in a CO_2 enriched atmosphere or by application of gibberelic acid, sodium chloride, urea or ammonium sulphate on stigmas. Alternatively, DH lines might be clonally propagated, in which case micro-propagation is often the best choice.

Mutation breeding is another area of plant improvement for which doubled haploid techniques can help to accelerate the process. Homozygosity of regenerants and true breeding propagation enables the fixation of mutations in the first generation after mutagenic treatment. All mutated traits are immediately expressed, allowing screening for both recessive and dominant mutants in the first generation without the need for self-pollination. The first option is, that mutagenic treatment is applied to dormant seeds that, on germination and flowering, produce M1 gametes, which are used as donor material for haploid culture. The second option relies on mutagenic treatment of haploid cells in vitro. The mutagenic agent is usually applied soon after microspore isolation at the uninucleate stage, before the first nuclear division in order to avoid heterozygosity and chimerism caused by spontaneous diploidization through nuclear fusion. In vitro mutagenic treatment can be followed by in vitro selection of desired traits, such as disease and herbicide resistance.

References

- Plant-Genetics: plantae.org, Retrieved 11, June 2020
- Plant-breeding, biology: byjus.com, Retrieved 09, February 2020
- Transgenic-plants-gene-constructs-vector-and-transformation-method, new-visions-in-plant-science: intechopen.com, Retrieved 14, August 2020
- Haploids-and-Doubled-Haploids-in-Plant-Breeding-221921962: researchgate.net, Retrieved 19, April 2020

Chapter 2

Plant Reproduction

Plant reproduction is the process by which plants generate new individuals or offspring. This can be accomplished through either sexual or asexual means. Sexual reproduction involves the fusion of gametes while asexual reproduction may occur through budding, fragmentation or spore formation. This chapter closely examines these types of plant reproduction for a better understanding of the topic.

Plant reproduction is the process by which plants generate new individuals, or offspring. Reproduction is either sexual or asexual. Sexual reproduction is the formation of offspring by the fusion of gametes. Asexual reproduction is the formation of offspring without the fusion of gametes. Sexual reproduction results in offspring genetically different from the parents. Asexual offspring are genetically identical except for mutation. In higher plants, offspring are packaged in a protective seed, which can be long lived and can disperse the offspring some distance from the parents. In flowering plants (angiosperms), the seed itself is contained inside a fruit, which may protect the developing seeds and aid in their dispersal.

Spores

Spores are reproductive cells that are able to develop into a new individual without fusing with another cell (in contrast, seeds are formed when male and female gametes join together). The microscopic spore cell has everything it needs to grow into a multicellular plant, and under favorable conditions the cell will divide and grow. In plants, spores can be found on non-seed bearing plants including green algae, mosses and ferns. Often, the spores are located on the underside of the leaves and are carried to a new area by wind or rain. Spores, unlike seeds, are less likely to be eaten by animals, but they are at risk of being consumed by bacteria and fungi.

Flowers

Flowers are what make flowering plants flowering plants! Flowers provide the mechanisms for sperm to find eggs, thus leading to fertilization and development of seeds.

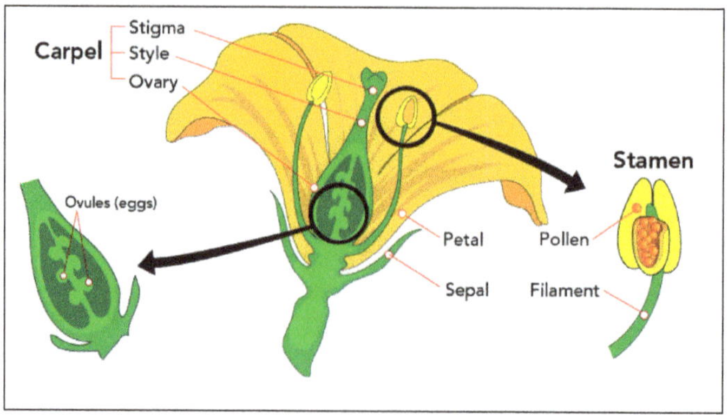

The outermost parts of the flower which typically surround the rest of the flower are green structures called sepals. Inside of the sepals are the petals, which are typically colourful. Petals on flowers are also modified leaves which serve a similar function as bracts. Next are the stamens, each of which contains a filament topped by pollen-producing cells. The innermost part of the flower is the carpel, which contains the ovary (where the eggs are located). The pollen from another flower must enter the ovary and fertilize the ovule in order for a seed to start developing.

Seeds

Seeds are embryonic (immature) plants that are enclosed in protective seed coats. Seeds often contain stored nutrients in the endosperm, which is rich in oil, starch and protein. Seeds can be dispersed (sent to new places) by wind (can be lighter or structured to be air-borne), by water (can float so they can drift down rivers) or by animals (can have barbs to catch on animal's fur or can be eaten and dispersed through an animal's droppings). Seeds are a more evolutionarily advanced form of plant reproduction than spores and are present in both gymnosperms and angiosperms. In gymnosperms the seeds are covered by the scales of cones, while in angiosperms the seed becomes covered with a fleshy or hard fruit.

Cones

Cones are the parts of conifers which contain reproductive structures. Female cones (A) produce ovules and male cones (B), which are much smaller and not as visible, produce pollen (which is visible as a yellowish powder).

The ovule, once fertilized by pollen, becomes a seed. In most conifer species, the male and female cones appear on the same plant, with the female cones on higher branches and the male cones on lower branches. This is to improve cross-fertilization (being fertilized with the pollen of another plant) and avoid self-fertilization (being fertilized with its own pollen). The tiny, lightweight pollen of one conifer is more likely to be carried by the wind to female cones of a different conifer than upward into its own female cones.

Fruit

Fruit is the result of the maturation of one or many flowers, and is therefore only found in angiosperms. In cooking, a fruit is any sweet-tasting plant product. However in plant science (botany), a fruit is considered the ripened ovary of a seed-bearing plant which contains its seeds. In flowering plants, as a seed develops, its ovary begins to ripen and the ovary wall, called the pericarp, becomes fleshy (as in berries and apples) or forms a hard outer covering (as in nuts). Botanically, fruit also include beans, corn kernels, tomatoes, wheat grains, pumpkins, peanuts, cucumbers and rice!

Plant Asexual Reproduction

Asexual reproduction is the formation of new individuals from the cell(s) of a single parent. It is very common in plants, less so in animals. All plant organs have been used for asexual reproduction, but stems are the most common.

Many plants are able to propagate themselves using asexual reproduction. This method does not require the investment required to produce a flower, attract pollinators, or find a means of seed dispersal. Asexual reproduction produces plants that are genetically identical to the parent plant because no mixing of male and female gametes takes place. Traditionally, these plants survive well under stable environmental conditions when compared with plants produced from sexual reproduction because they carry genes identical to those of their parents.

Figure: Different types of stems allow for asexual reproduction.

(a)The corm of a garlic plant looks similar to (b) a tulip bulb, but the corm is solid tissue, while the bulb consists of layers of modified leaves that surround an underground stem. Both corms and bulbs can self-propagate, giving rise to new plants. (c) Ginger forms masses of stems called rhizomes that can give rise to multiple plants. (d) Potato plants form fleshy stem tubers. Each eye in the stem tuber can give rise to a new plant. (e) Strawberry plants form stolons: stems that grow at the soil surface or just below ground and can give rise to new plants.

Many different types of roots exhibit asexual reproduction. The corm is used by gladiolus and garlic. Bulbs, such as a scaly bulb in lilies and a tunicate bulb in daffodils, are other common examples. A potato is a stem tuber, while parsnip propagates from a taproot. Ginger and iris produce rhizomes, while ivy uses an adventitious root (a root arising from a plant part other than the main or primary root), and the strawberry plant has a stolon, which is also called a runner.

Some plants can produce seeds without fertilization. Either the ovule or part of the ovary, which is diploid in nature, gives rise to a new seed. This method of reproduction is known as apomixis.

An advantage of asexual reproduction is that the resulting plant will reach maturity faster. Since the new plant is arising from an adult plant or plant parts, it will also be sturdier than a seedling. Asexual reproduction can take place by natural or artificial (assisted by humans) means.

Vegetative Reproduction

Propagation is the natural mechanism by which plants regenerate. Propagation by seeds is the main method by which plants reproduce in nature. The method is also efficient and widely used in tree propagation. As a matter of fact, all trees produced seeds, except those individuals bearing only male flowers in some species. The process of seed production begins during flowering; when the pollen is transferred from the male part (anther) to the female part (stigma) of the flower-seed being the end product of a long, complex and physiological process.

In certain plant families such as Burceraceae, Clusiaceae and Euphobiaceae, it is common to find dioecious plants (separate male and female trees). This implies some plant (male trees) have functional stamens with anthers (male part) in their flowers while others (female trees) have functional pistile with stigmas (female parts) in the flowers. Examples include; Allanblackia spp, Canarium schwweinfurthii (Black fruit) and Ricinodendron heudelotii (njangsang) in which only trees bearing female flowers produce fruit and ultimately seeds. It is often very difficult to distinguish between male and female trees except when they are in flower. Another phenomenon common in some plant families such as Sterculiaceae, Burceraceae is hermaphroditism in which case plants will possess male and hermaphrodite flowers. The two types of flowers occur mixed on the same inflorescences, though occasionally on different inflorescences or trees as in Cola nitida and Cola acuminata and in Dacryodes edulis. Studies on the floral biology of Cola for example has indicated incompatibility in the fertilisation of hermaphrodite flowers with pollen from male flowers from within or between trees exist. Thus in the propagation of such species by seeds, at least 5 seeds should be planted to increase the chances of having a female tree. However, the surest way of obtaining either a female or male tee is through vegetative propagation.

Vegetative Propagation of Agroforestry Trees

Vegetative propagation of plants is essentially the reproduction of plant material from vegetative organs so that the offspring will contain the exact characteristics of the parent plant with regards to genotypes and health status. This is possible because plants unlike animals have meristimatic, undifferentiated cells that can differentiate to the various organs necessary to form a whole new plant. Thus a piece of plant shoot, root, or leaf can grow to form a new plant that contains the exact genetic information of its source plant.

While sexual reproduction has been responsible for the wide genetic diversity observed in many tropical tree species as a result of constant recombination of genes, vegetative propagation aims at the identical reproduction of plants with desirable features such as high productivity, superior quality, resistance, high tolerance to biotic and abiotic stress and therefore plays an important role in continuing preferred traits over generations. Vegetative propagation has been used in fruit tree improvement since biblical times, and continues to still relevant today in agroforestry tree domestication efforts.

Methods of Vegetative Propagation

Cuttings are severed uninodal leafy shoot or root fragments that can develop adventitious roots and shoots respectively. Uninodal leafy shoot cuttings are usually place into a suitable rooting substrate and kept under high humidity in propagators until roots and shoots are formed. Propagation by cuttings is a relatively easy method and can provide a large number of propagules in a very short time. Cuttings have been widely used in clonal propagation of forest trees where large numbers of planting stocks are required and recently in the production of quality germplasm of local fruit trees (Irvingia gabonesis, Ricinodendron hedelotii, Dacryodes edulis, Allanblackia spp), medicinal plants (Prunus africana, Pausinystallia johimbe).

Grafting is a technique used to untie 'parts' of different plants by bring the cambium of each into contact and then creating a situation under which the cut surfaces can unite and grow away together. There are many grafting methods, some very specific for a particular species. They depend on size and shape of the scions and the place on the rootstock onto which they are grafted; they can be carried out at various times throughout the year. The technique involves two important stages: the preparation of the grafting surfaces and the procedures for aftercare. In addition, the technique also requires grafting skill first, and then extended experience to become familiar with the aftercare procedure without which newly grafted plants might be lost due to poor aftercare.

Scion - The part of the graft that will provide the new system. The scion maybe united either at the apex or side of the rootstock.

Rootstock – The lower part of the graft. It normally possesses a root system that will support the subsequent shoot development from the scion.

Layering (marcotting) technique is similar to cuttings with the advantage that the propagule is detached from the parent plant only after development of roots. Its multiplication rate is lower than that of cuttings though the plants are larger. Methods of layering are the oldest used for vegetative propagation. Three 'classes' of layering methods, depending on the soil and stem relationship can be distinguished:

- Soil is mounded to the stem (mound layering, trench layering).
- Stems are lowered into the soil (simple layering, tip layering).
- Stems are not covered by soil (air layering).

Air-layering is the technique that has been widely used in the domestication of indigenous fruit trees in west and central Africa in capturing the attributes of elite trees within genetically diverse wild populations, so avoiding the long, slow process of tree breeding.

Micropropagation is a specialized method of propagation in which small pieces of plant tissues are regenerated on an artificial medium under sterile conditions. It embraces the regeneration from shoot and root tips, callus tissue, leaves, seed embryo, anthers and even single cells. Micropagation has following the applications in woody plant multiplication.

- As a means to remove viruses from plants, thereby improving clones of plants in clonal improvement programmes.
- As a very effective and rapid method to multiply high quality clonal material (bulking-up of plants).

Micropapagtion has its limitations and should not be regarded as the answer to all propagation problems. Some limitations are summarized below:

- The initial capital investment in the laboratory and equipment can be high.
- Micropropagation requires skill, usually specially trained laboratory manager and skilled disciplined support staff. All these entail additional cost.

Notwithstanding, Micropropagation should be seen as a technology to use in conjunction with conventional propagation techniques. Macropropagation is therefore, a "tool" to refine the production of some high value crops, to produce healthier plant material.

Rationale for Propagating Agroforestry Trees Vegetatively

Maintaining Superior Genotypes

Most tropical agroforestry tree species are out breeders – Allanblackia spp, Canarium schwweinfurthii (Black fruit) and Ricinodendron heudelotii (njangsang) Cola spp; implying that through the recombination of genes during sexual reproduction, many important characteristics might disappear. If a superior individual tree/ideotype has been identified by farmers or researchers, its genetic information can be 'fixed' through vegetative propagation, thus allowing the reproduction of the same superior individual in the next generation.

Problematic Seed Germination and Storage

Some tree species produce seedless fruits (e.g. off-season Dacryodes edulis,) and need to be propagated vegetatively, others bear fruit very scarcely or erratically (Prunus africana); and some bear fruits abundantly but due to some form of physiological dormancy germination is very slow and the rates are quiet low Allanblackia spp,. Many tropical tree species have recalcitrant seeds that require special and often cumbersome seed handling procedures. In these cases, vegetative propagation might be a suitable and cheaper alternative to seedling production.

Shortening Time to Flower and Fruit

An important reason for vegetative propagation is the shortening of the reproductive cycle of a tree. This is particularly important when the flowers, fruits or seeds are the desired products. Most vegetative propagation is done with scions or cuttings from mature trees, which maintain the characteristics of maturity after grafting or rooting as will be explained above. For instance Irvingia gabonesis, Dacryodes edulis, Ricinodendron heudelotii and Cola spp marcotts will fruit from 3 years after field planting. Grafts of Irvingia spp, Garcinia kola and Allanblackia have been observed to fruit in nursery 6 months after grafting.

Combining More than one Genotype in one Plant

Grafting is a unique way of combining desired characteristics from two or more plants into a single one. Scions with particular fruit characteristics can be grafted onto rootstocks with other desirable characteristics such as disease resistance and adaptability to environmental constraints. Another possibility is the grafting of more than one cultivar or species onto the same stem, for example Irvingia gabonensis (sweet fruits) grafted to an Irvingia wombolu (bitter fruits) rootstock or to extend the period of bearing by grafting early and late varieties on a single tree. Sometimes, it could even be the introduction of a pollinator branch onto a female tree in the case of dioecious species like Ricinodendron heudelottii, Canarium schwweinfurthii and Allanblackia spp.

Uniformity of Plantations

For many commercially grown species - Citurs spp, Mangifera indica, uniformity of growth form or fruiting season is important economically. This phenomenon will sooner or later be relevant with the increasing demand for agroforestry tree products - Irvingia kernel, Dacryodes edulis fruits, Cola nuts and Allanblackia seed.

Application of Vegetative Propagation in Domestication of Agroforestry Trees

Many tree species are dioecious. Farmers are more interested in the female, fruit-bearing individuals and need only a few male pollinator trees around. It is difficult if not impossible to predict in small seedlings, whether they are female or male, and often trees change their sex expression with age. In these cases, vegetative propagation of the female trees could help increase the number of fruit bearing trees on farm.

Selection (Capture of Desirable Traits)

Layering was developed, by the Chinese 4-5000 years ago. Although the layering techniques are quite simple, their commercial use today is restricted to specific applications. The reason for this is that the productivity of layers is far smaller than that of cuttings. However, the use of air layering technique has been widely used in the domestication of indigenous fruit trees in west and central Africa.

Grafting has become the most used way to improve high-valued fruit trees like citrus, mango, or avocado. This is most likely attributed to the opportunities they offer to combine different traits in the different plant parts. ICRAF-WCA/HT is currently researching on the application of grafting techniques in the improvement of economically important local fruit trees of west and central Africa. Grafting has successfully been adapted to Irvingia gabonesis (bush mango), Ricinodendron heudelotii (Njangsang), Garcina kola (bitter cola), and Allanblackia spp with grafted plants promising to fruit under 5 years as opposed to at least 10 years in the wild.

Despite the potential of grafting in the domestication of agroforestry trees, it also has some limitations. For example, graft incompatibilities arise when rootstock and scion do not readily join. Therefore, in order to produce successful and sustainable grafts, compatible clones have to be identified. This is a resource-consuming process that is inevitable for the success of the technique. Also, grafting techniques demand some skill.

Mass Propagation

Vegetative propagation by cuttings of several local fruit trees and medicinal plants has received considerable, attention in the last decade from ICRAF-WCA/HT research programme on domestication of agroforestry trees. Propagation by cuttings, in low-cost 'non-mist' poly-propagators made of local material is rapidly being adopted by ICRAF partners, (NARS, NGOs and farmers) in Cameroon, Nigeria, Equatorial Guinea, Gabon, Democratic Republic of Congo and Ghana and Tanzania.

Tree Improvement

Farmers often report that when a tasty or large fruit from a superior fruit tree for example is planted, the seedling does not perform up to their expectations. Scientific reasons advanced to explain these observations stems from the fact that, the pollinator of the seeds was a different fruit tree though of same species and thus the seeds have a different gene combination than the mother tree. Therefore, the ability to produce large and tasty fruits was not inherited. On the other hand, the application of vegetative propagation techniques can provide an exact copy of the mother tree from which the seed was harvested and retain the desired fruit quality in the following generation if the traits are heritable.

Vegetative propagation of Allanblackia species Methods have been developed for the vegetative propagation of Allanblackia floribunda using single-node leafy stem cuttings in non-mist propagators (Atangana et al., in press). Factors identified to influence the rooting of cuttings are rooting medium, genetic origin (clone) and presence/absence of leaf. Best rooting percentages (68.7%) were obtained in leafy cuttings from one clone inserted in sand. Experiments are also in progress for development of grafting protocols in Allanblackia. Preliminary results indicate that Allanblackia gabonensis trees are amenable to grafting, the best practice being side-tongue graft (80% success). Specific issues in Allanblackia vegetative propagation A. floribunda is a hard-to-root species, as cuttings develop roots at least 10 weeks after insertion in the propagator. Best rooting percentages are obtained after 24 weeks, what constitute an issue to overcome before scaling-up protocols of propagation. As a bonus, genetic origin of cuttings (clone) strongly influences the rooting ability in Allanblackia cuttings, indicating that the ability to clone particular trees will be an important factor in the selection that should be done following the ideotype concept, using ideotypes combining fruit and nut traits, as well as ability to root.

Protocol of Propagation of Allanblackia using Rooting of Cuttings

- Choice of cuttings: As one of the objectives is to massively propagate plus trees, an option would be to harvest cuttings from recently felled trees known to produce fruits with desired characteristics (high productivity, heavy fruits with many seeds...). Juvenile cuttings should be harvested from coppicing stumps, sprayed with water and stored in moist polythene bags (to reduce the stress) and brought to the nursery.
- Preparation of cuttings: Using a sharp blade, pest and disease-free shoots are cut into single-node cuttings of 3-4 cm length. The leaf lamina of each cutting should be trimmed to 12.5, 25 or 50 cm2 with scissors.
- Non-mist propagators: High-humidity, non-mist propagators consisting of a wooden frame enclosed in clear polythene sheets so that the base of the propagator is watertight (3 m x 1 m x 1 m), and the lid is hermetically sealed, should be used. Sand should be the rooting medium, treated with fungicide (Ridomil Plus (metalaxyl-M and cupper) and Caocobre (cupper): 50 g each dissolved in 15 l of water) and insecticide (Cyperdim 220 EC 200g/l dimethoate and 20g/l cypermethrine: 40 ml dissolved in 1.5 l of water) 3 days prior to the beginning of the experiment.
- Insertion of cuttings in the propagators and follow-up: Three-hundred cuttings can be inserted in one non-mist propagator. Caocobre can be applied to the rooting medium when necessary (fungal attacks) during the rooting process. Each morning (very early), water level in the propagator should be monitored, and corrections made when excess or deficit. Cuttings should then be sprayed with a fine jet of water. Assessments of rooting success should start 10 weeks after cuttings' insertion in the propagator, by lifting the cuttings from the rooting medium. A cutting is considered to be rooted when it has one or more roots exceeding 1 cm. Rooted cuttings should be removed from the propagator a potted in black polythene bags (1 l) containing a 2:1 mixture of forest soil and sand.

  Cuttings are then brought to a giant propagator for weaning (2 weeks). After 2 weeks, cuttings are removed from the giant propagator and keep in the nursery (shade) before transplantation that can occur 6 months later (cuttings should be transplanted in a shade place, otherwise, their leaves will be burnt by sun).

Protocol for Allanblackia Grafting

Materials you need for grafting:

- Pruning shears (pruners).
- Budding – knife or well sharpened knife.

- Whetstone or fire.
- Fastening film.
- Protective plastic.

Once the material is available, grafting of Allanblackia can be made using the 'sidetongue' grafting which has proven to give the best success rate (80%) in Allanblackia gabonensis, as follows:

- Choice of scion: The choice of a scion is a very important step through grafting of Allanblackia, as grafting consists in uniting a selected tree with a rootstock. Hence, scions should be harvested from trees with known characteristics (female trees with heavy and many seeded fruit).
- Preparing the scion:
    - Gather (15-20cm long) 6-12 months old branches that can produce buds.
    - Choose a scion whose diameter matches that of the rootstock.
    - Trim the base of the branches with a bevel.
- Preparing the rootstock:
    - Choose a plant, with size of a pencil about 8mm.

The plant diameter will depend on the growing speed of the species which can reach this size anytime between 3 and 12 months.

- Remove the leaves and branches over 30 cm above the collar.
- Make a side cut of 2-3 cm.

- Joining the rootstock and the scion:
    - Insert the trimmed base of scion in the cut of the rootstock adjusting the edges of the scion to those of the rootstock.
    - Fasten the graft from bottom to top with fastening film.
    - Protect the scion entirely with a transparent plastic.

Cleft grafting could also be used, as it has the advantage of still grafting unsuccessfully grafted plants using the side-tongue graft, as the entire shoot of the plant is cut off between 10 and 30 cm before grafting with scion. Cleft grafting is done as follows:

Preparation of the Rootstock

- Eliminate leaves and branches over 15-20cm above the collar when the rootstock has size of pencil (8mm).
- Cut the rootstock 15 or 20cm above the collar.

- Make an incision of 2-3cm from the apex.

Preparation of the Scion

- Cut branches of 15-20cm long and matches that during buds.
- Choose a scion whose diameter matches that of the rootstock.
- Trim the base of the scion with a bevel.

Joining the Rootstock and Scion

- Insert the trimmed base of the scion in the incision of the rootstock adjusting the edges of the scion to those of the rootstock.
- Fasten the graft from bottom to top.
- Cover the scion with a transparent plastic.

Taking care of Grafts

- Undo the plastic once the bud cover develops (14 days after the grating) on the scion.
- Gradually cut the rootstock each week until it is 1cm from the ligature.
- Cut the fastening film when the shoot has 7 leaves.
- Water, fertilize, weed and treat with a fungicide or an insecticide.
- Eliminate all other shoots on the rootstock other that the scion.

Precautions

- Graft as soon as the branches are taken.
- Avoid damaging the plants accidentally with grafting equipment.
- The cut part must be clean.
- Secure the ligature tightly to ensure good contact.
- Use sharp instruments and disinfect them with spirit or fire.
- Repeat the grafting operation if the scion or shield bud dies (two weeks after grafting).

Protocol for Allanblackia Marcotting

Marcotting is a form of vegetative propagation that consists of inducing rooting of branches still attached on a tree. After rooting the branch is cut and transferred to a nursery to enable it to grow leaves and become an independent plant. Like all other vegetative propagation technique marcotting allow for the reproduction of exact copies of the mother tree and for early fructification as compared to seedling plant. On the three main existing classes or marcotting, air layering is the one currently being experienced in A. floribunda. It consists of inducing the rooting on a branch of a tree having desirably characteristics a farmer will want to reproduce. After rooting, the branch is cut off, potted and placed in a humidity chamber in the nursery.

Air layering is done as follows:

a) Selection, Placing and Monitoring of Marcotts

- Select your best tree (good, taste, off-season tree, big-fruit yield tree).
- During April, May and June climb up the trees and choose suitable orthotropic branches with a diameter between 3 to 5 cm with the aid of a knife, strip the bark over a length of between 5 cm to 10 cm (choose a reasonable number of branches 10 maximum such that after rooting, the tree would not be over trimmed).
- Scratch the stripped part of the branch to remove the cambium (the slippery part) if not it will reconstitute the bark.
- Sheath the wound with well decomposed sawdust or moist infloresence of palm.
- The sheath is made with transparent polythylene paper so that roots can be seen when it develop.
- If there are many ants and squirrels in the area, check marcotts regulary (once every, month) and adjust those sheath that are about to be destroyed by ants and squirrels.

b) Harvesting

- Roots generally appear depending upon the species two to four months after the marcott is set. Cut the rooted marcotts to a length of 40 cm to 50 cm.
- Label and moisture the marcott and make sure that it remains permanently fresh from the place of harvest to the nursery.

c) Potting Rooted Marcotts

- Prepare a substrate composed of 1/3 sand and 2/3 black soil or comprising a mixture of 1/3 sand 1/3 black soil, 1/3 compost.
- Dip the stripped marcott in a basin of water.
- Free the roots by carefully removing the substrate between them.
- Insert the marcotts into pots with a lot of care avoid damaging the roots.
- With bee-wax, wood-ash, or mastic, cover the wounded part of the marcott to limit infections.
- Water the marcott and place it in the humidity chamber.

- Water it regularly without flooding it.
- When buds appear, remove the marcott immediately and place it under slight shade with temperature and moisture similar to the outside environment to permit it develop and harden-off.
- One to two months later, the marcotts can be transplanted once the leaves are fully develop.
- Once the marcott is on the farm, it can be given any other phytosanitary treatment and care as other plants.

Apomixis

Apomixis is defined as "The process of development of diploid embryos without fertilization." Or "Apomixis is a form of asexual reproduction that occurs via seeds, in which embryos develop without fertilization." Or "Apomixis is a type of reproduction in which sexual organs of related structures take part but seeds are formed without union of gametes."

In some species of plants, an embryo develops from the diploid cells of the seed and not as a result of fertilization between ovule and pollen. This type of reproduction is known as apomixis and the seedlings produced in this manner are known as apomicts. Apomictic seedlings are identical to mother plant and similar to plants raised by other vegetative means, because such plants have the same genetic make-up as that of the mother plant. Such seedlings are completely free from viruses.

In apomictic species, sexual reproduction is either suppressed or absent. When sexual reproduction also occurs, the apomixes is termed as facultative apomixis. But when sexual reproduction is absent, it is referred to as obligate apomixis.

Apomixis was defined by Winkler as replacement of the normal sexual reproduction by asexual reproduction, without fertilization. This definition notable does not mention meiosis. Thus "normal asexual reproduction" of plants, such as propagation from cuttings or leaves, has never been considered apomixis but replacement of the seed by a plantlet, or replacement of the flower by bulbils are types of apomixis. Apomixis produces embryos or seed progeny that are exact genetic replicas of the mother plant. The main advantage of apomixis over sexual reproduction is the possibility to select individual plants with superior characteristics and propagate them clonally through seeds.

Types of Apomixis

a. Recurrent Apomixis

In this type of apomixis, the embryo sac develops from the diploid egg mother cell or from some other diploid cells of the embryo sac without fertilization. Consequently, the egg has the normal diploid number of chromosomes, the same as in the mother plant. The embryo subsequently develops from egg nucleus without fertilization. This series of events is known to occur in some species of Crepis, Poa Taraxacum and Allium without the stimulus of pollination, while in others, like Parthenium, Rubus, Malus (apple) the stimulus of pollination appears to be necessary for the development embryo or to produce a viable endosperm.

b. Non-Recurrent Apomixis

In this type of apomixis, the embryo develops directly from the haploid egg cell (haploid parthenogenesis) or some other haploid cells of embryo sac (haploid apogamy) without fertilization, and as a result, the embryo developed is also haploid. This type of apomixis occurs very rarely and is primarily of genetic interest. The plants produced are haploid and sterile. It is common in Solanum nigrum and Lilium species etc.

c. Adventitious Embryony or Nucellar Embryony

Adventitious embryony is also called as nucellar embryony or nucellar budding. In this type, the embryo does not develop from the cells of the embryo sac but it develops from a cell or a group of cells either of nucellus or integuments. The cells of integument or nucellus are diploid and hence the resultant embryo is also diploid. Such embryos usually develop outside the embryo sac in addition to the regular embryo. Adventitious embryony occurs in many plant species but is most prominent in tropical and sub-tropical tree species, like citrus and mango.

d. Vegetative Apomixis or Bulbils

In this type, vegetative buds or bulbils are produced in the inflorescence in place of flowers. These buds or bulbils may sprout into new plants while they are still attached to the mother plants. This type of apomixis is quite common in Allum, Agave, Poa and Diascorea and in some grasses.

Advantages

- Assured reproduction in the absence of pollinators, such as in extreme environments.
- Maternal energy not wasted in unfit offspring (cost of meiosis).
- Some apomictic plants (but not all) avoid the male energy cost of producing pollen.

Disadvantages

- Cannot control accumulation of deleterious genetic mutations.
- Usually restricted to narrow ecological niches Lack ability to adapt to changing environments.

Monoembryony

"It is the phenomenon in which only one embryo produced in a seed." POLYEMBRYONY "It is the phenomenon in which two or more embryos are produced in a seed." Polyembryony means that more than one embryo develops within a single seed. It is also known as adventitious embryony (Nucellar embryony or Nucellar budding). This condition may result from many reasons but one of the most common reasons being the nucellar embryony. Sometimes more than one nucleus develops within the embryo sac, which may lead to development of more than one embryo. Further, cleavage of pro-embryo during early stages of development may be other reason for development of multiple embryos as in case of conifers. Among fruit crops polyembryony is quite common in mango and citrus.

- Polyembryony can develop from several distinct causes. Specific cells in the nucellus or sometimes with integument have embryos. Genetically, these embryos have the same genotype as the parental plant and are apomictic.
- Adventitious embryony occurs in many plant species but is most common in citrus and mango. In these species, both zygotic and apomictic embryos are produced. In other species (e.g. Opuntia), no pollination or fertilization is needed.
- Polyembryony is common in mango and citrus. In trifoliate orange (Poincirus trifoliate) several seedlings arise from one seed. Of these seedlings, one seedling, usually the weakest may be sexual, and the others arise apomictically from cells in the nucellus, which are diploid copies of the mother plant.

Horticultural Significance of Polyembryony

Nucellar seedlings in citrus are completely free from viruses, because the embryo sac and adjoining tissues are impregnated at flowering time with some unknown powerful substances which kills all the viruses. For immediate requirement of planting material, development of nucellar lines is the quickest and easiest method. The major possible horticultural applications of polyembryony are:

- Nucellar seedlings are true-to-type seedlings.
- Such seedlings are genetically uniform and can be used as virus free rootstocks.
- More vigorous seedlings – continuous vegetative propagation leads to decline in vigour in citrus.
- Development of virus free seedlings and bud wood.
- Significance in breeding programme.

Significance of Polyembryony and Apomixis

Polyembryony and apomixis occur in many plant species in nature. The ability of plants to reproduce apomictically has great significance in horticulture because of the following reasons:

- These provide an opportunity for producing genetically uniform population.
- Apomictic seedlings are useful for producing uniform rootstocks in apple (Malus toringoides and M. sikkimensis), citrus and Jamun etc.
- Polyembryony demonstrates that embryogenic potential is not limited to zygote but is possessed by various other somatic cells, which forms the base for in vitro somatic embryogenesis.
- Apomictic seedlings are highly uniform and quite healthy in contrast to sexually produced seedlings.

- Apomictic seedlings are helpful for identification during hybridization (e.g. trifoliate leaf character in citrus).
- Apomictic seedlings are suitable for screening viruses, which are not otherwise transmitted through seeds.

Developmental Features of Apomixis—Three Essential Components

Apomeiosis and Adventitious Embryony

Apomixis can be broadly divided into the following three types: adventitious embryony, a sporophytic type, or diplospory and apospory, two gametophytic (apomeiotic) types. There is considerable variation in the timing of developmental events within an ovule undergoing apomictic and sometimes concurrent amphimictic processes. This variation has led to the classification of numerous subtypes of gametophytic apomixis.

Sporophytic differs from gametophytic apomixis in that no alternation of generations intervenes prior to embryo development. The embryo forms directly from a somatic cell in the ovule, typically a nucellar, but sometimes an integumentary cell. Since seed development also requires a nutritive endosperm, and endosperm can only develop during the gametophytic phase of the life cycle, amphimixis is concurrent with apomixis in sporophytic apomicts. The simultaneous occurrence of adventitious embryos and reduced embryo sacs allows for endosperm to develop after fertilization of the central cell in the reduced embryo sac. The endosperm can thus nourish either an adventitious embryo or the zygotic embryo when one out-competes the other, or both may develop to relative maturity resulting in polyembryony.

Gametophytic apomixis can completely replace amphimixis in an "obligate" apomict, although it is unlikely that completely obligate apomicts actually exist. Within the ovule of a gametophytic apomict, the megaspore mother cell is either arrested in its development or does not enter or complete meiosis. Without meiosis, haploid megaspores cannot develop and all cells of the unfertilized ovule retain the somatic chromosome number. Diplospory results when the megaspore mother cell or one of its diploid daughter cells develops into an unreduced female gametophyte. Considerable variation in megaspore mother cell development prior to diplosporous embryo sac initiation can occur and has led to the classification of multiple types of diplospory. The most common are the Antennaria type, also known as mitotic diplospory where the megaspore mother cell directly undergoes mitosis, and the Taraxacum type with a restitution of the somatic chromosome number at meiosis I. In apospory, the origin of the unreduced female gametophyte is a somatic cell, leading to the possibility that both amphimictic and apomictic processes could occur in the same ovule in facultative apomicts. Such situations are relatively rare, however, and one process usually out-competes the other. The competitiveness of sexual versus apomictic development probably depends in part on heterochronic shifts in reproductive events. The plasticity of these events within and between apomictic genotypes has been documented in Tripsacum accessions and hybrids with maize as well Hieracium genotypes. A similar conclusion was reached from study of these diplosporous and aposporous taxa—apomixis is most likely a deregulation of the sexual process. However, examples are known where the absence of apomeiotic events in an ovule of an apomict does not restore successful sexual reproduction, suggesting that there is a specific factor required at least for initiation of apomeiosis if not for the downstream events of embryo sac development.

Parthenogenesis

Parthenogenesis is the development of an unfertilized gamete into a plant. In amphimicts as well as apomicts, two sperm are released from the pollen tube that penetrates a synergid in an embryo sac; however, the fate of the sperm in an apomict differs from that in an amphimict. Typically, no sperm fertilizes the egg because it develops parthenogenetically. In exceptional cases, an unreduced egg can be fertilized resulting in a hybrid embryo with an elevated ploidy level. When meiosis occurs in an apomict, allowing chromosome reduction, it typically is followed by syngamy between egg and sperm; however, the egg may occasionally develop parthenogenetically to produce a haploid (or polyhaploid) embryo. Possible nuclear states of embryo and endosperm are shown in Table below.

Table: Combinations of reduced (Red) or unreduced (NonRed) nuclei due to fertilization (Fert), or lack of recombination due to autonomous development (Parth), that can lead to different ploidy states in embryo or endosperm cells. The first or nuclear state derives from the egg (embryo) or polar nuclei (endosperm) while the subsequent ones are of sperm origin.

	Red/Fert	NonRed/Fert	Red/Parth	NonRed/Parth
Embryo	n+1 n+2n	2n + n 2n +2n	n+0	2n + 0
Endosperm	[n+n] +n [n+n] + 2n	[2n +2n] + n [2n + 2n] +2n [2n + 2n] + n+n 2n + n	[n+n]+0	[2n +2n] + 0

Embryo development from an unreduced egg can be precocious in an apomict. Precocity is particularly pronounced in Tripsacum dactyloides. Interestingly, out of 5534 transcriptional units surveyed with maize microarrays, no gene expression differences were observed between unfertilized maize ovules and unfertilized, apomictic maize-Tripsacum hybrid ovules even though the embryo had begun to develop prior to pollination in the apomictic hybrid. A significant difference in the proportion of polyadenylated mRNA was found between the two genotypes, however, suggesting that mRNA maturation may play a greater role in embryo initiation than transcription, and that the maternal to zygotic transition, in terms of altered transcriptional profile, does not occur until after the initial zygotic divisions.

Even in Pennisetum ciliare where parthenogenetic embryo development is not considered to be precocious, ultrastructural examination revealed a complete cell wall around the egg cell of apomictic P. ciliare prior to penetration of the embryo sac by a pollen tube. This pattern differed in the chromosomally reduced egg where exposed membrane remained at the chalazal end. Precocious cell wall development in the apomeiotic egg may prevent fertilization. Indeed, more 2n + n hybrids were recovered when pollination took place 2 to 3 days prior to anthesis instead of at anthesis.

One interesting model for parthenogenesis in plants is the "Salmon" system in wheat. "Salmon" contains a wheat-rye translocation in its nucleus that, when in a particular cytoplasmic background, will result in male sterility and haploid parthenogenesis. While the Salmon system is not apomictic, it does provide useful isogenic lines for molecular studies including the construction of egg-cell-specific cDNA libraries.

Fertilization and Endosperm Development

The central cell must be fertilized in many apomicts in order for endosperm to develop. This phenomenon is known as pseudogamy. Some apomicts, e.g., many in the Asteraceae, do not require fertilization of the central cell and the endosperm develops autonomously. The maternal to paternal genome contribution to the endosperm and its relationship to the ploidy level of the embryo have a strong effect on seed viability in amphimicts. Maize has been particularly well studied in this regard and was shown to have normal endosperm development only when the maternal to paternal genome ratio in the endosperm was 2m: 1p. Apomicts either are more tolerant of variation in maternal: paternal genome ratios or have adaptations that restore such ratios. For example, apomictic genotypes of Tripsacum dactyloides produce [2n (4x) +2n (4x)] +1n (2x) endosperm, i.e., two unreduced polar nuclei fuse with a reduced sperm to generate an 8m: 2p ratio in normally developing endosperm. An adaptation to avoid endosperm imbalance is displayed by apomictic Pennisetum species as well as other panicoid apomicts that develop four-nucleate embryo sacs. Fournucleate embryo sacs typically have a uninucleate central cell that is fertilized to create 2n + n endosperm or reconstitute a 2m: 1p genome ratio. A change in this developmental pattern occurs when apomixis is introgressed into sexual P. glaucum from an apomictic relative: the frequency of uninucleate central cells drops dramatically in favor of binucleate central cells and is correlated with a reduction in seed set.

Apomictic Genomes

Evolution/Phylogenetic Distribution

Apomixis is widely distributed among angiosperm families where both sporophytic and gametophytic types have been reported among monocots, eudicots, and magnoliids. Apomixis clearly originated multiple times during flowering plant evolution. It appears to be most widespread in the grass (Poaceae) and sunflower (Asteraceae) families although its frequency could be due, in part, to the intensity with which reproduction has been studied in these families, both of which contain crop species. However, Solanaceae also contains many crop species (e.g., tomato, potato, tobacco), but apomixis is relatively infrequent. Both sporophytic and gametophytic types of apomixis often are present in the same plant family as are the diplosporic and aposporic forms of gametophytic apomixis. The occurrence of multiple types of apomixis in a family, genus, species, and sometimes individual genotype could be interpreted as a result of overlapping or intersecting developmental pathways that are subject to genotype x environment effects. Evidence is accumulating that apomixis likely is a consequence of deregulation of sexual reproduction, although the point of deregulation can vary and may be responsible for the different developmental phenotypes.

Since apomixis and amphimixis are not mutually exclusive, genetic variation can be found in agamic complexes and often is greater than would be expected by extrapolating from the frequency of aposporous or diplosporous embryo sac development. Apomixis is not a tightly controlled developmental process, but is subject to considerable plasticity with regard to fate of an egg cell and its coordinate development with the endosperm. All of the egg cell cytological fates shown in Table are possible in an apomictic individual, particularly where residual sexual reproduction has been documented. Similarly, potential cytological states in the endosperm are very variable when apomixis as a whole is considered, but usually are more limited within any individual apomict. An example of reproductive diversity comes from Hypericum perforatum where the flow cytometric seed screen was used to document 11 different reproductive outcomes of embryo and endosperm combinations. This extreme plasticity may be unusual, but definitely could contribute to the genetic diversity observed within or among populations or agamic complexes. Progeny from crosses between Australian populations of H. perforatum had an average of 2.5% meiotic embryo sac development, which, although seemingly low, would allow for numerous recombination events to occur over multiple generations. Whether variation in populations of apomicts is due primarily to recombination or mutation has been debated, although recent evidence from Hieracium pilosella, using a method of cladistics analysis called component compatibility analysis, implicated residual sexual reproduction as the primary determinant of genotypic diversity.

Genetic and Genomic Analysis

Sporophytic Apomixis

Sporophytic apomixis has been reported to be inherited simply as a dominant trait. In the only mapping study published thus far on apomictic citrus, a 3:1 segregation ratio was observed for apomixis in a cross of Citrus volkameriana X Poncirus trifoliata. However, QTL mapping with molecular markers and phenotyping for nucellar embryony and polyembryony resulted in the identification of six QTL for both positive and negative effects, a more complex pattern than anticipated. Further genetic and molecular studies in Citrus and its relatives, as well as other species reproducing by adventitious embryony should be forthcoming.

Aposporous Apomixis In most species, apospory appears to be simply inherited based on segregation of the trait in crosses between aposporous male parents and strictly sexual female parents. Some of these crosses have been made between biotypes within a species and others between different sexually compatible species. Since apomicts are always highly heterozygous, segregation of the trait for mode of reproduction as well as molecular markers can be observed in the F1 generation. This strategy differs from most inheritance and genetic mapping studies where inbred lines exhibiting different traits would be crossed. In some studies, both the male (apomict) and female (amphimict) maps have been constructed using data from the F1 population. Apospory rarely has been shown to segregate from parthenogenesis and behaves as a dominant trait, inherited in Mendelian fashion, although sometimes subject to segregation distortion. This pattern of inheritance has been observed for Pennisetum squamulatum; Cenchrus ciliaris syn. Pennisetum ciliare, Panicum maximum, Brachiaria sp., Paspalum notatum, Ranunculus sp., and Hieracium sp. The only clearly documented case of recombination between apospory and parthenogenesis is found inPoa pratensis. InPoa, parthenogenesis can be readily phenotyped in the absence of fertilization by the development of embryos after auxin treatment. Using auxin treatment, parthenogenesis had been mapped as a single gene, qualitative trait using AFLP (amplified fragment length polymorphism) and SCAR (sequence characterized amplified region) markers. From a similar cross of sexual and apomictic genotypes, a small number of non-parthenogenetic genotypes were observed to produce aposporous embryo sacs. The functionality of these embryo sacs was not tested. A more complex genetic model was postulated for Poa pratensis that included single, unlinked genes for apospory initiation, apospory prevention, parthenogenesis initiation, and parthenogenesis prevention as well as a megaspore development gene. This model was predicated on the evolution of apomixis from sexual plants and was supported by the observation of discrete classes of expressivity that could best be explained by the five-gene model.

As with parthenogenesis in Poa, apospory typically is phenotyped as a qualitative trait (i.e., an individual either has the capacity to produce aposporous embryo sacs or only meiotic embryo sacs), and not quantitatively with respect to the frequency with which aposporous embryo sacs can either be formed or function. Environment is known to affect frequency of mode of reproduction in some species, but not others, which suggests that modifiers can play a role in the penetrance of the trait in some species; however, none of these modifiers have been mapped to date. Other evidence for modifiers comes from the reproductive phenotype of backcrosses involving an apomictic trispecific hybrid of Pennisetum glaucum, P. purpureum, and P. squamulatum. A BC3 plant produced >99% maternal progenies, but several BC4 plants produced only up to 89% maternal progenies. The difference in penetrance of the trait is most likely due to modifiers because both generations carried an intact chromosome from P. squamulatum that was shown to be the only genomic region required for transfer of apomixis. Modifiers may also play a role in fate of embryo sac nuclei since the majority of central cells in BC3 were binucleate compared with uninucleate central cells in P.squamulatum.

A single regulatory gene has been proposed as sufficient for the induction of apomixis, and although simple genetic inheritance appears to support this hypothesis, molecular evidence suggests that more complex genetic control of the entire apomixis process cannot be discounted. In particular, the linkage groups typically transmitted with apospory display large blocks of non-recombining molecular markers leading to speculation that adapted gene complexes within supernumerary chromatin might be required for the function of at least certain types of apomixis. The association of apomixis with a non-recombining region of the genome first was described in P. squamulatum. Extensive characterization of this chromosomal region using RFLPs and fluorescence in situ hybridization of apomixis-linked bacterial artificial chromosome clones has shown the region to be extremely large in size (estimated at least 50 Mbp), heterochromatic, and highly hemizygous. A heterochromatic and hemizygous region is found in P. squamulatum as well as C. ciliaris, both polyploid apomicts, and is indicative of heteromorphism between the homologous chromosomal pairing partners apparently due to an insertion in both species combined with an inversion/translocation in P. squamulatum. Lack of recombination and homozygosity have not been confined to apomictic Pennisetum and close relatives, but also were found in Paspalum simplex. The association of apospory with a heterochromatic region of the genome, rich in retro transposons, as shown for Pennisetum, raises the intriguing possibility that chromatin structure and/or RNA interference could play a role in control of "apomixis" gene expression.

Diplosporous Apomixis

Contrary to apospory, linkage between diplosporous embryo sac development and parthenogenesis can readily be broken in at least two taxa of the Asteraceae, Erigeron and Taraxacum. In crosses of sexual diploid and apomictic triploid Taraxacum genotypes, many 3x and 4x hybrids set seed in the absence of pollination as would be expected for apomictic Taraxacum where the endosperm develops autonomously. However, several 3x hybrids did not set seed when left unpollinated, although one produced exclusively 2n + n embryos when pollinated with a diploid genotype. This genotype was incapable of parthenogenesis. The independence between diplospory and parthenogenesis was further confirmed in a diplosporous 4x × sexual 2x cross, which produced only 5x progeny, and the DIP allele was shown to be linked to the 18S-25S rDNA locus in the reciprocal cross. Autonomous endosperm development also has been shown to segregate independently from parthenogenesis.

Similarly, progeny from a cross of sexual diploid and apomictic triploid genotypes of Erigeron showed a range of chromosome numbers that, except for the diploid state, were not strongly predictive of mode of reproduction. Nevertheless, a single locus model for diplospory was a close fit to the data. While all parthenogenic plants were diplosporous, several diplosporous plants did not form embryos suggesting that a genetic component for parthenogenesis had been eliminated. When one such diplosporous, but nonparthenogenic triploid plant, was used as a female parent in crosses with a sexual diploid male parent, predominantly tetraploid (2n + n) progeny were recovered. Genetic mapping of the segregating F1 population provided 4 and 12 AFLP markers linked to parthenogenesis and diplospory, respectively. Clearly, the two traits were segregating independently. Markers for parthenogenesis were not in a region of low recombination.

One diplosporous grass species, Tripsacum dactyloides, has been extensively studied, in particular because of its potential for transfer of apomixis to maize through traditional breeding. Diplospory was determined to be inherited simply in a cross of Zea mays × Tripsacum dactyloides, and five RFLP markers that had been mapped to the long arm of maize chromosome 6 were found to be strictly co-segregating with diplospory. However, in a sexual, diploid Tripsacum cross, considerable recombination between the five linked markers was observed. The localization of apomixis to a maize-Tripsacum chromosome translocation also supported the conclusion that only a single Tripsacum chromosome transmitted apomixis.

The association of apospory and diplospory with re-combinationally suppressed chromosomal regions has now been observed in aposporous (Pennisetum, Paspalum) and diplosporous (Erigeron, Tripsacum) species. Surprisingly, the pattern of low recombination has been broken upon construction of an AFLP map of the diplospory region in Taraxacum. Since diplosporous embryo sac development was inherited as a single, dominant gene (DIP), bulked segregant analysis was used to identify 34 AFLP markers that were linked and spanned a distance of 18.6 cM, although none was found to strictly co-segregate with diplospory. Taraxacum provides a singular example of recombination in a genomic region harboring genes for apomeiotic embryo sac formation.

Model Systems for the Study of Apomixis Aided by Comparative Genomics

The overwhelming majority of apomicts is not crop species and/or has relatively large genome sizes; therefore, molecular tools for their direct analysis are not well developed. There are some criteria that are desirable for both model sexual and apomictic species (•), although other criteria are specific to apomicts (◦):

- Easy to cultivate.
- Easy to propagate:
    - Perennial (for maintenance of specific genotypes).

- Compact plant stature.
- Short generation time.
- Abundant seed set.
- Small genome size:
    - Easy to cytologically evaluate mode of reproduction.
    - Easy to genetically evaluate mode of reproduction.
    - Highly apomictic (i.e., low level of residual sexual reproduction).
    - Fertile pollen (required for genetic studies).
    - Cross-compatible sexual and apomictic biotypes in the same species, preferably at the same ploidy level.
- Diploid (extremely rare for apomicts).
- Transformation competent.

All of the monocot apomicts are relatively large in plant stature and are likely to have low transformation efficiencies. However, since monocots and eudicots are evolutionarily distant, it would be pertinent to study apomixis in both groups of plants. Comparative genetic mapping is one method to transfer information from model species to larger genome species that have fewer molecular tools. Although genome sequence of one eudicot and one monocot is now available, the most direct application would be to closely related species. In this respect, the Boechera holboellii (syn. Arabis holboellii) complex has the greatest advantage since it is in the same family (Brassicaceae) as the first plant genome sequenced Arabidopsis thaliana. This apomictic species complex is diplosporous and primarily pseudogamous. Apomictic Boechera are often triploid (2n = 21), but can also be a neuploid or diploid, although the most recent analyses indicate that some B. holboellii apomicts are near-diploid aneuploids (2n = 15) containing a B chromosome rather than strict diploids. Whether the B chromosome is always associated with the apomixis phenotype remains to be tested.

Apomictic grasses also have the advantage of one sequenced genome in the family, that of Oryza sativa. Comparative mapping in cereals has shown that some linkage blocks can be highly conserved in gene order and content across species that diverged from one another approximately 60 million years ago. This macro-synteny observed on a genetic map scale, however, often can be interrupted by translocation of chromosomal regions from another part of the genome or intra- and inter-chromosomal duplications. In apomicts, comparative mapping has been a target of several projects in order to utilize the genome information from model species such as rice and maize. The results have not demonstrated common regions of synteny associated with apomixis among multiple apomictic grass species. Grimanelli et al. argue that such results may indicate that apomixis has evolved independently in related species and may have a different genetic basis in each. It is not unprecedented that a single phenotype, dwarf being an example, can result from the mutation of positionally unrelated genes, although the genes may operate through the same metabolic or developmental pathway.

Comparative mapping in apomictic plants largely has been carried out with restriction fragment length polymorphism (RFLP) analysis. Early studies identified common RFLP fragments among multiple apomictic Pennisetum species/genotypes. However, anonymous DNA clones from P. squamulatum that were linked with apomixis often did not hybridize to sexual Pennisetum species, including pearl millet (P. glaucum); therefore, they were not useful for comparative mapping with other grasses. More recently, limited DNA sequencing of BAC clones localized to a region of the P. squamulatum genome shown to be transmitted with apospory has allowed the identification of likely orthologous regions in rice based on sequence similarities (Conner et al., in preparation). Macrosynteny between P. squamulatum and rice appears to be highly interrupted although regions of micro-synteny are apparent. Other mapping efforts in apomictic grasses have taken advantage of available cDNA clones mapped in sexual species such as maize and rice. Using cDNA clones mapped in rice on a segregating population of Paspalum simplex, apospory was localized to a region of the genome syntenic with the long-arm telomeric region of rice chromosome 12. Synteny within this region was complete between P. simplex and P. malacophyllum but partial with P. notatum. The apomixis-associated chromosome in P. notatum appears to have been involved in a translocation, and as a consequence, the region containing two of the five mapped rice markers falls on another chromosome in this species and can be eliminated from consideration as containing a major effect gene for apomixis.

Ploidy Variation

Genetic and genomic analysis of apomixis is further complicated by the polyploid nature of essentially all apomicts. Polyploidy does not appear to be required for apomixis to be expressed. There are multiple examples of polyhaploid plants that have developed from parthenogenetic reduced eggs in apomicts, and such plants, although often lacking vigor, nevertheless reproduce by apomixis. One example where expression of apomixis may be dependent on ploidy level or gene dosage, perhaps not of the apospory factor itself but of an epistatic regulator, comes from Paspalum. Certain diploid genotypes of Paspalum form vigorous stands with no sympatric occurrence of tetraploids and therefore are unlikely to have arisen as dihaploids. Nevertheless, aposporous embryo sacs can be found at a low frequency in these genotypes. This frequency increases dramatically when the chromosomes of the diploids are doubled. Aposporous embryo sacs were not observed in an induced tetraploid from a strictly sexual diploid genotype, suggesting that penetrance of the apomixis phenotype was dependent on ploidy level.

In several studies, transmission of apomixis "genes" through haploid gametes appears to have been impaired, probably as a result of linked deleterious mutations associated with a typically non-recombining region of the genome or of the "apomixis factor" itself. An exception is found in Hieracium where haploid gametes were shown to be viable in a cross of sexual 4x × apomictic 3x genotypes since many apomictic triploids were recovered from n + n events; however, the diploid zygote apparently was disadvantaged in sexual 2x × apomictic 3x seeds because no apomictic diploids were produced. In addition to the occasional parthenogenetic development of a reduced egg of an apomict, apomixis predicts that higher ploidy levels also would arise through the occasional fertilization of unreduced eggs with either reduced or unreduced sperm. The varied combinations of reduction/non-reduction with fertilization/parthenogenesis result in a range of possible ploidy levels and participate in the formation of agamic complexes through diploidtetraploid-dihaploid cycles, e.g., Bothriochloa-Dichanthium.

Gene Expression in Ovules of Apomicts

In addition to structural analysis of a natural apomict genome, several groups have sought to isolate genes differentially expressed between ovules of an apomictic versus sexual genotype. Much of this work has been implemented with differential display, which uses arbitrary primers. One caveat of differential display stems from the highly heterozygous and thus polymorphic at the DNA sequence level, nature of all apomicts. It is not possible to create isogenic lines that vary only in their mode of reproduction; therefore, any expressed gene identified by nucleotide sequence-based techniques would have to be verified, preferably by northern blot analysis or sequencing of corresponding cDNA clones, as differing in expression rather than just DNA sequence polymorphism. In some cases, the effect of this heterozygosity has been minimized by using pools of mRNA from different individuals, one pool from apomictic individuals and the second from sexual.

Differential display has successfully detected bands uniquely amplified in either apomictic or sexual genotypes or their respective pools. In two studies (Brachiaria and Pennisetum), mature ovules were used for RNA extraction and although differential expression was validated in one study, the cloned genes are unlikely to be responsible for the induction of apomixis. In similar work with Paspalum notatum, verification of differentially expressed transcripts was only carried out by RT-PCR, a technique that also would be affected by sequence polymorphisms. A cDNA library screen was used to isolate a single cDNA from Panicum maximum that was expressed in flowers of the apomict but not the sexual. Differential display performed on mature embryo sac stage. A simplistic model of heterochronic changes in gene expression (pink) between amphimictic and apomictic ovules, which follows Nogler's postulate that "the physiological conditions for the activation of genes involved in the differentiation of the embryo sac, which normally prevail only in megaspores, prevail also in certain unreduced cells." An embryo sac initiation gene could be misregulated in apomicts such that earlier expression in the megaspore mother cell could prevent meiosis and lead directly to embryo sac development by mitosis. Similarly, ectopic expression in somatic nucellar cells could result in the development of aposporous embryo sacs. The resulting embryo sacs may be indistinguishable among developmental types. Developing embryo and endosperm stage ovules of Hieracium resulted in the isolation of a DEFICIENS homolog that showed transient down-regulation in the chalazal region of ovules from apomicts. Recently, careful staging of ovary development has led to the identification of differentially expressed transcripts in Brachiaria and Poa. It is likely that careful staging will be critical for interpretation of results, particularly if misexpression, rather than unique gene expression, is responsible for the switch in mode of reproduction.

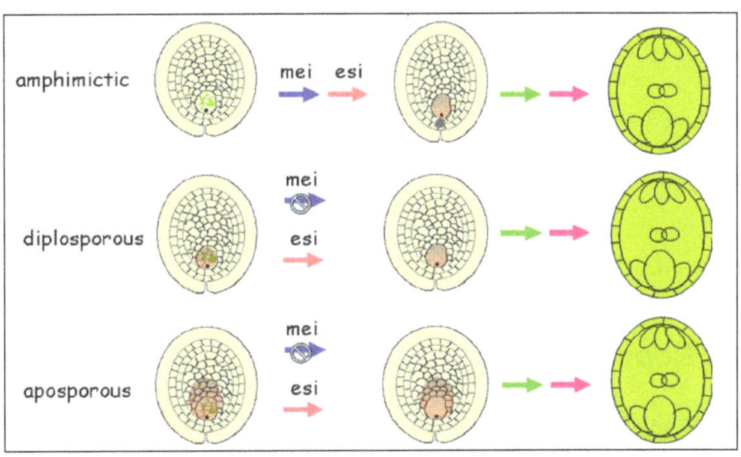

A simplistic model of heterochronic changes in gene expression (pink) between amphimictic and apomictic ovules, which follows Nogler's postulate that "the physiological conditions for the activation of genes involved in the differentiation of the embryo sac, which normally prevail only in megaspores, prevail also in certain unreduced cells." An embryo sac initiation gene could be misregulated in apomicts such that earlier expression in the megaspore mother cell could prevent meiosis and lead directly to embryo sac development by mitosis. Similarly, ectopic expression in somatic nucellar cells could result in the development of aposporous embryo sacs. The resulting embryo sacs may be indistinguishable among developmental types.

Perhaps the most thorough study using a transcript profiling approach comes from Poa pratensis where cDNA-AFLP analysis resulted in the isolation of fragments that were specific to carefully staged florets of either a sexual or apomictic genotype and were not present in leaves. Out of 2248 transcript-derived fragments, 179 were qualitatively or quantitatively different between the sexual and apomictic genotypes. After sequencing and BLAST analysis, the majority of fragments showed similarity to genes expressed in reproductive organs of other species. Of eight full-length cDNA sequences cloned (PpRAB1-1, PpRAB1-2, PpARM, PpAPK, PpSERK1, PpSERK2, APOSTART1, and APOSTART2), none was specifically expressed in apomictic or sexual genotypes but rather their expression was differentially modulated or quantitatively different, lending additional support to the hypothesis that apomixis results from a deregulated sexual pathway.

Plant Sexual Reproduction

In sexual reproduction, two parents are involved in producing a new individual. Offspring is produced by the fusion of gametes (sex cells) from each parent. Animals like dog, cats, lions, giraffe, humans, etc., all reproduce sexually.

Figure: Sexual Reproduction in Plants.

Process of Sexual Reproduction

This fundamental process of reproduction is DNA replication. In other words, if an organism wants to continue its species, then they need to transfer their traits to the next generation by DNA (genetic information) copying which occurs during the process of reproduction.

We know asexually reproducing organisms copy their DNA and divide themselves into new cells which are their clones. Since the process of replication is not completely reliable, there are chances of variation, but they are quite limited. Organisms need to adapt themselves to the changing environment; the creation of variants needs to be hastened for this reason. This will only happen when there is a union of two different DNA's. This highlights the significance of sexual reproduction in organisms.

If two reproducing cells simply join, the chromosome number in the new organism will be doubled. But this cannot happen because the number of chromosomes in each cell of every organism is fixed – the same holds good for the reproducing cell. Thus, the reproducing cells from both parents reduce their chromosome number into the half before fusion and are called gametes (germ cells). In simple organisms, the size and shape of gametes are almost the same. But in higher organisms, a male gamete is small and mobile while female gamete is large, immobile and they store food for the fetus. The gametes are produced in a specialized system called the reproductive system.

Features of Sexual Reproduction

- Two parents are involved (both male and a female).
- Gamete formation and fertilization take place.
- The whole process is slow and lengthy.
- The variation occurs; offspring are different from parents, genetically and physically.

Reproduction in plants takes place sexually and asexually as well. But the majority of the flowering plants reproduce sexually. The flower is the reproductive part of a plant i.e., both male and female gametes are produced by flowers. Sexual reproduction in plants takes place in flowers. The complete flower typically consists of four parts:

- Petals.
- Sepals.
- Stamen (male reproductive part).
- Pistil/Carpel (female reproductive part).

Stamen (male reproductive part) consists of anther and filament.

- The anther is a sac-like structure that produces and stores pollen.
- The filament supports the anther.

The pistil (female reproductive part) comprises three parts- stigma, style, and ovary.

- Stigma is the topmost part of a flower.

- The style is the long tube which connects the stigma to the ovary.
- The ovary contains a lot of ovules. It is the part of the plant where the seed formation takes place.

A flower may consist of either stamen or pistil or both. Based on this, a flower can be either uni sexual or bisexual. A bisexual flower is composed of all the four parts e.g. Rose, China rose. Whereas, plants like papaya and cucumber produce only uni sexual flowers.

Pollination and Fertilization

In order to form a zygote, male gametes in pollen grains have to fuse with egg in the ovule. This is achieved by the process called pollination. Pollination is the process of transferring pollen grains from the anther – male part of a flower, to the stigma – female part of a flower. Depending on the pollen landing, pollination can be classified into two types.

Self-Pollination and Cross-Pollination

In angiosperms, pollination is defined as the placement or transfer of pollen from the anther to the stigma of the same flower or another flower. In gymnosperms, pollination involves pollen transfer from the male cone to the female cone. Upon transfer, the pollen germinates to form the pollen tube and the sperm for fertilizing the egg. Pollination has been well studied since the time of Gregor Mendel. Mendel successfully carried out self- as well as cross-pollination in garden peas while studying how characteristics were passed on from one generation to the next. Today's crops are a result of plant breeding, which employs artificial selection to produce the present-day cultivars. A case in point is today's corn, which is a result of years of breeding that started with its ancestor, teosinte. The teosinte that the ancient Mayans originally began cultivating had tiny seeds—vastly different from today's relatively giant ears of corn. Interestingly, though these two plants appear to be entirely different, the genetic difference between them is miniscule.

Pollination takes two forms: self-pollination and cross-pollination. Self-pollination occurs when the pollen from the anther is deposited on the stigma of the same flower, or another flower on the same plant. Cross-pollination is the transfer of pollen from the anther of one flower to the stigma of another flower on a different individual of the same species. Self-pollination occurs in flowers where the stamen and carpel mature at the same time, and are positioned so that the pollen can land on the flower's stigma. This method of pollination does not require an investment from the plant to provide nectar and pollen as food for pollinators.

Living species are designed to ensure survival of their progeny; those that fail become extinct. Genetic diversity is therefore required so that in changing environmental or stress conditions, some of the progeny can survive. Self-pollination leads to the production of plants with less genetic diversity, since genetic material from the same plant is used to form gametes, and eventually, the zygote. In contrast, cross-pollination—or out-crossing—leads to greater genetic diversity because the micro gametophyte and mega gametophyte are derived from different plants.

Because cross-pollination allows for more genetic diversity, plants have developed many ways to avoid self-pollination. In some species, the pollen and the ovary mature at different times. These flowers make self-pollination nearly impossible. By the time pollen matures and has been shed, the stigma of this flower is mature and can only be pollinated by pollen from another flower. Some flowers have developed physical features that prevent self-pollination. The primrose is one such flower. Primroses have evolved two flower types with differences in anther and stigma length: the pin-eyed flower has anthers positioned at the pollen tube's halfway point, and the thrum-eyed flower's stigma is likewise located at the halfway point. Insects easily cross-pollinate while seeking the nectar at the bottom of the pollen tube. This phenomenon is also known as heterostyly. Many plants, such as cucumber, have male and female flowers located on different parts of the plant, thus making self-pollination difficult. In yet other species, the male and female flowers are borne on different plants (dioecious). All of these are barriers to self-pollination; therefore, the plants depend on pollinators to transfer pollen. The majority of pollinators are biotic agents such as insects (like bees, flies, and butterflies), bats, birds, and other animals. Other plant species are pollinated by abiotic agents, such as wind and water.

Methods of Pollination

Pollination by Insects

Bees are perhaps the most important pollinator of many garden plants and most commercial fruit trees. The most common species of bees are bumblebees and honeybees. Since bees cannot see the color red, bee-pollinated flowers usually have shades of blue, yellow, or other colors. Bees collect energy-rich pollen or nectar for their survival and energy needs. They visit flowers that are open during the day, are brightly colored, have a strong aroma or scent, and have a tubular shape, typically with the presence of a nectar guide. A nectar guide includes regions on the flower petals that are visible only to bees, and not to humans; it helps to guide bees to the center of the flower, thus making the pollination process more efficient. The pollen sticks to the bees' fuzzy hair, and when the bee visits another flower, some of the pollen is transferred to the second flower. Recently, there have been many reports about the declining population of honeybees. Many flowers will remain unpollinated and not bear seed if honeybees disappear. The impact on commercial fruit growers could be devastating.

Figure: Insects, such as bees, are important agents of pollination.

Many flies are attracted to flowers that have a decaying smell or an odor of rotting flesh. These flowers, which produce nectar, usually have dull colors, such as brown or purple. They are found on the corpse flower or voodoo lily (Amorphophallus), dragon arum (Dracunculus), and carrion flower (Stapleia, Rafflesia). The nectar provides energy, whereas the pollen provides protein. Wasps are also important insect pollinators, and pollinate many species of figs.

Figure: A corn earworm sips nectar from a night-blooming Gaura plant.

The pollen is picked up and carried on the butterfly's limbs. Moths, on the other hand, pollinate flowers during the late afternoon and night. The flowers pollinated by moths are pale or white and are flat, enabling the moths to land. One well-studied example of a moth-pollinated plant is the yucca plant, which is pollinated by the yucca moth. The shape of the flower and moth have adapted in such a way as to allow successful pollination. The moth deposits pollen on the sticky stigma for fertilization to occur later. The female moth also deposits eggs into the ovary. As the eggs develop into larvae, they obtain food from the flower and developing seeds. Thus, both the insect and flower benefit from each other in this symbiotic relationship. The corn earworm moth and Gaura plant have a similar relationship.

Pollination by Bats

In the tropics and deserts, bats are often the pollinators of nocturnal flowers such as agave, guava, and morning glory. The flowers are usually large and white or pale-colored; thus, they can be distinguished from the dark surroundings at night. The flowers have a strong, fruity, or musky fragrance and produce large amounts of nectar. They are naturally large and wide-mouthed to accommodate the head of the bat. As the bats seek the nectar, their faces and heads become covered with pollen, which is then transferred to the next flower.

Pollination by Birds

Many species of small birds, such as the hummingbird and sun birds, are pollinators for plants such as orchids and other wildflowers. Flowers visited by birds are usually sturdy and are oriented in such a way as to allow the birds to stay near the flower without getting their wings entangled in the nearby flowers. The flower typically has a curved, tubular shape, which allows access for the bird's beak. Brightly colored, odorless flowers that are open during the day are pollinated by birds. As a bird seeks energy-rich nectar, pollen is deposited on the bird's head and neck and is then transferred to the next flower it visits. Botanists have been known to determine the range of extinct plants by collecting and identifying pollen from 200-year-old bird specimens from the same site.

Figure: Hummingbirds have adaptations that allow them to reach the nectar of certain tubular flowers.

Pollination by Wind

Most species of conifers, and many angiosperms, such as grasses, maples and oaks, are pollinated by wind. Pine cones are brown and unscented, while the flowers of wind-pollinated angiosperm species are usually green, small, may have small or no petals, and produce large amounts of pollen.

Figure: A person knocks pollen from a pine tree.

Unlike the typical insect-pollinated flowers, flowers adapted to pollination by wind do not produce nectar or scent. In wind-pollinated species, the microsporangia hang out of the flower, and, as the wind blows, the lightweight pollen is carried with it.

The flowers usually emerge early in the spring, before the leaves, so that the leaves do not block the movement of the wind. The pollen is deposited on the exposed feathery stigma of the flower.

Figure: These male (a) and female (b) catkins are from the goat willow tree (Salix caprea). Note how both structures are light and feathery to better disperse and catch the wind-blown pollen.

Pollination by Water

Some weeds, such as Australian sea grass and pond weeds, are pollinated by water. The pollen floats on water, and when it comes into contact with the flower, it is deposited inside the flower.

Pollination by Deception

Orchids are highly valued flowers, with many rare varieties. They grow in a range of specific habitats, mainly in the tropics of Asia, South America, and Central America. At least 25,000 species of orchids have been identified.

Flowers often attract pollinators with food rewards, in the form of nectar. However, some species of orchid are an exception to this standard: they have evolved different ways to attract the desired pollinators. They use a method known as food deception, in which bright colors and perfumes are offered, but no food. Anacamptis morio, commonly known as the green-winged orchid, bears bright purple flowers and emits a strong scent. The bumblebee, its main pollinator, is attracted to the flower because of the strong scent—which usually indicates food for a bee—and in the process, picks up the pollen to be transported to another flower.

Figure: Certain orchids use food deception or sexual deception to attract pollinators. Shown here is a bee orchid (Ophrys apifera).

Other orchids use sexual deception. Chiloglottis trapeziformis emits a compound that smells the same as the pheromone emitted by a female wasp to attract male wasps. The male wasp is attracted to the scent, lands on the orchid flower, and in the process, transfers pollen. Some orchids, like the Australian hammer orchid, use scent as well as visual trickery in yet another sexual deception strategy to attract wasps. The flower of this orchid mimics the appearance of a female wasp and emits a pheromone. The male wasp tries to mate with what appears to be a female wasp, and in the process, picks up pollen, which it then transfers to the next counterfeit mate.

Double Fertilization

After pollen is deposited on the stigma, it must germinate and grow through the style to reach the ovule. The microspores, or the pollen, contain two cells: the pollen tube cell and the generative cell. The pollen tube cell grows into a pollen tube through which the generative cell travels. The germination of the pollen tube requires water, oxygen, and certain chemical signals. As it travels through the style to reach the embryo sac, the pollen tube's growth is supported by the tissues of the style. In the meantime, if the generative cell has not already split into two cells, it now divides to form two sperm cells. The pollen tube is guided by the chemicals secreted by the synergids present in the embryo sac, and it enters the ovule sac through the micropyle. Of the two sperm cells, one sperm fertilizes the egg cell, forming a diploid zygote; the other sperm fuses with the two polar nuclei, forming a triploid cell that develops into the endosperm. Together, these two fertilization events in angiosperms are known as double fertilization. After fertilization is complete, no other sperm can enter. The fertilized ovule forms the seed, whereas the tissues of the ovary become the fruit, usually enveloping the seed.

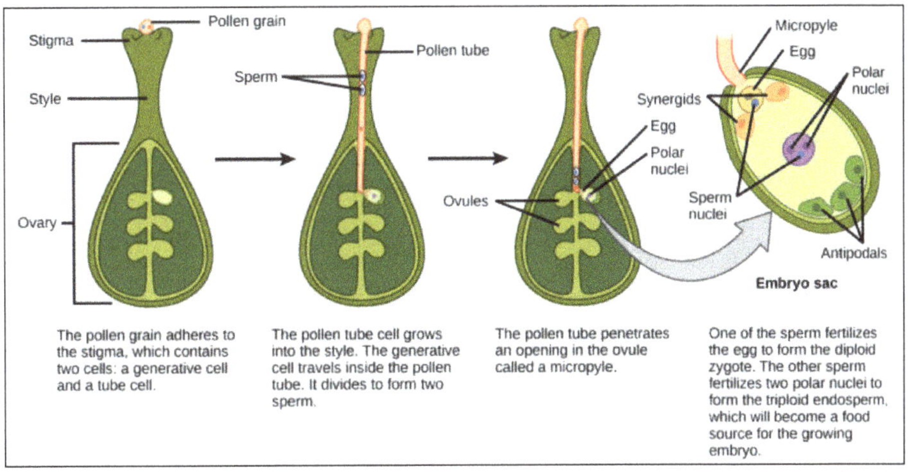

Figure: In angiosperms, one sperm fertilizes the egg to form the 2n zygote, and the other sperm fertilizes the central cell to form the 3n endosperm. This is called a double fertilization.

After fertilization, the zygote divides to form two cells: the upper cell, or terminal cell, and the lower, or basal, cell. The division of the basal cell gives rise to the suspensor, which eventually makes connection with the maternal tissue. The suspensor provides a route for nutrition to be transported from the mother plant to the growing embryo. The terminal cell also divides, giving rise to a globular-shaped proembryo. In dicots (eudicots), the developing embryo has a heart shape, due to the presence of the two rudimentary cotyledons. In non-endospermic dicots, such as Capsella bursa, the endosperm develops initially, but is then digested, and the food reserves are moved into the two cotyledons. As the embryo and cotyledons enlarge, they run out of room inside the developing seed, and are forced to bend. Ultimately, the embryo and cotyledons fill the seed, and the seed is ready for dispersal. Embryonic development is suspended after some time, and growth is resumed only when the seed germinates. The developing seedling will rely on the food reserves stored in the cotyledons until the first set of leaves begin photosynthesis.

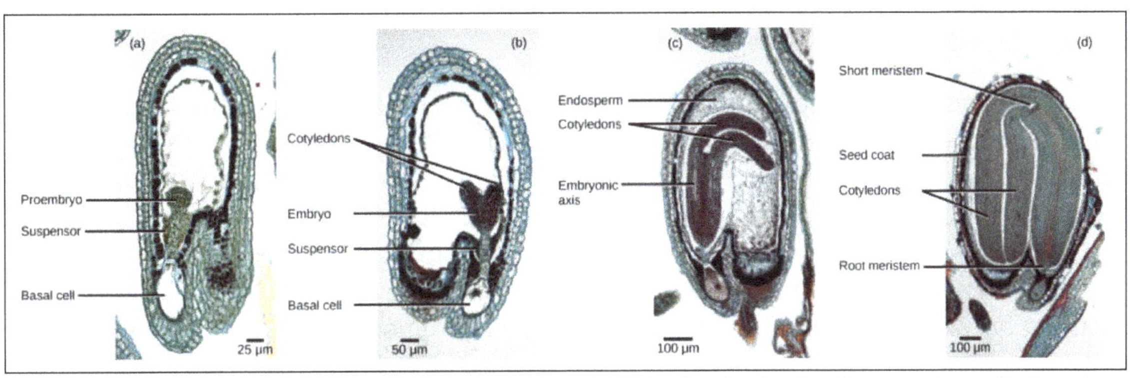

Figure: Shown are the stages of embryo development in the ovule of a shepherd's purse (Capsella bursa).

After fertilization, the zygote divides to form an upper terminal cell and a lower basal cell. (a) In the first stage of development, the terminal cell divides, forming a globular pro-embryo. The basal cell also divides, giving rise to the suspensor. (b) In the second stage, the developing embryo has a heart shape due to the presence of cotyledons. (c) In the third stage, the growing embryo runs out of room and starts to bend. (d) Eventually, it completely fills the seed.

Development of a Seed

The mature ovule develops into the seed. A typical seed contains a seed coat, cotyledons, endosperm, and a single embryo.

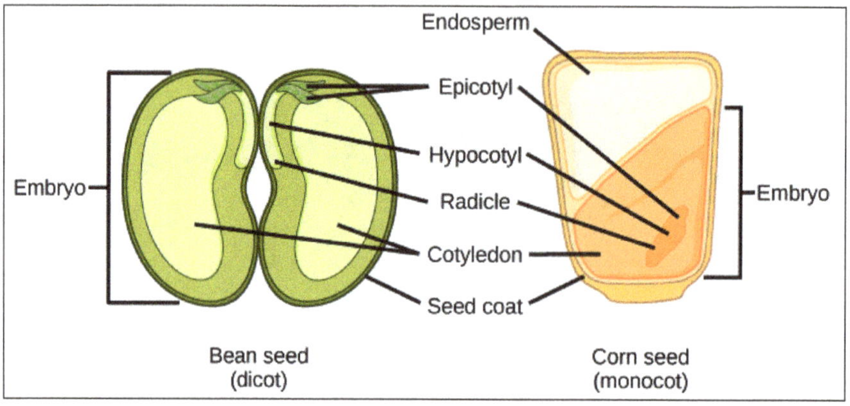

Figure: The structures of dicot and monocot seeds are shown.

Dicots (left) have two cotyledons. Monocots, such as corn (right), have one cotyledon, called the scutellum; it channels nutrition to the growing embryo. Both monocot and dicot embryos have a plumule that forms the leaves, a hypocotyl that forms the stem, and a radicle that forms the root. The embryonic axis comprises everything between the plumule and the radicle, not including the cotyledon(s).

The storage of food reserves in angiosperm seeds differs between monocots and dicots. In monocots, such as corn and wheat, the single cotyledon is called a scutellum; the scutellum is connected directly to the embryo via vascular tissue (xylem and phloem). Food reserves are stored in the large endosperm. Upon germination, enzymes are secreted by the aleurone, a single layer of cells just inside the seed coat that surrounds the endosperm and embryo. The enzymes degrade the stored carbohydrates, proteins and lipids, the products of which are absorbed by the scutellum and transported via a vasculature strand to the developing embryo. Therefore, the scutellum can be seen to be an absorptive organ, not a storage organ.

The two cotyledons in the dicot seed also have vascular connections to the embryo. In endospermic dicots, the food reserves are stored in the endosperm. During germination, the two cotyledons therefore act as absorptive organs to take up the enzymatically released food reserves, much like in monocots (monocots, by definition, also have endospermic seeds). Tobacco (Nicotiana tabaccum), tomato (Solanum lycopersicum), and pepper (Capsicum annuum) are examples of endospermic dicots. In non-endospermic dicots, the triploid endosperm develops normally following double fertilization, but the endosperm food reserves are quickly remobilized and moved into the developing cotyledon for storage. The two halves of a peanut seed (Arachis hypogaea) and the split peas (Pisum sativum) of split pea soup are individual cotyledons loaded with food reserves.

The seed, along with the ovule, is protected by a seed coat that is formed from the integuments of the ovule sac. In dicots, the seed coat is further divided into an outer coat known as the testa and inner coat known as the tegmen.

The embryonic axis consists of three parts: the plumule, the radicle, and the hypocotyl. The portion of the embryo between the cotyledon attachment point and the radicle is known as the hypocotyl (hypocotyl means "below the cotyledons"). The embryonic axis terminates in a radicle (the embryonic root), which is the region from which the root will develop. In dicots, the hypocotyls extend above ground, giving rise to the stem of the plant. In monocots, the hypocotyl does not show above ground because monocots do not exhibit stem elongation. The part of the embryonic axis that projects above the cotyledons is known as the epicotyl. The plumule is composed of the epicotyl, young leaves, and the shoot apical meristem.

Upon germination in dicot seeds, the epicotyl is shaped like a hook with the plumule pointing downwards. This shape is called the plumule hook, and it persists as long as germination proceeds in the dark. Therefore, as the epicotyl pushes through the tough and abrasive soil, the plumule is protected from damage. Upon exposure to light, the hypocotyl hook straightens out, the young foliage leaves face the sun and expand, and the epicotyl continues to elongate. During this time, the radicle is also growing and producing the primary root. As it grows downward to form the tap root, lateral roots branch off to all sides, producing the typical dicot tap root system.

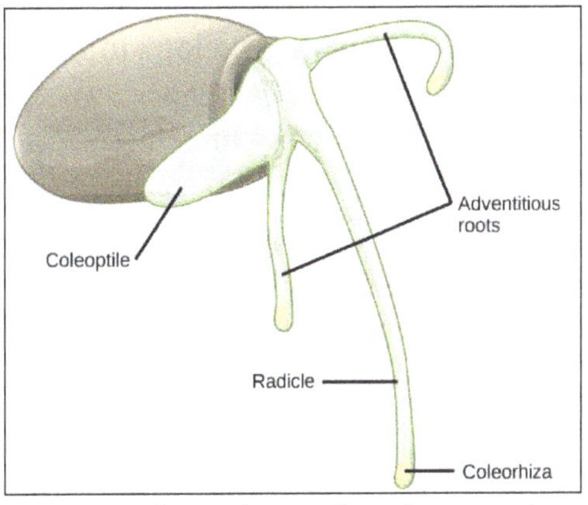

Figure: As this monocot grass seed germinates, the primary root, or radicle, emerges first, followed by the primary shoot, or coleoptile, and the adventitious roots.

In monocot seeds, the testa and tegmen of the seed coat are fused. As the seed germinates, the primary root emerges, protected by the root-tip covering: the coleorhiza. Next, the primary shoot emerges, protected by the coleoptile: the covering of the shoot tip. Upon exposure to light (i.e. when the plumule has exited the soil and the protective coleoptile is no longer needed), elongation of the coleoptile ceases and the leaves expand and unfold. At the other end of the embryonic axis, the primary root soon dies, while other, adventitious roots (roots that do not arise from the usual place – i.e. the root) emerge from the base of the stem. This gives the monocot a fibrous root system.

Seed Germination

Many mature seeds enter a period of inactivity, or extremely low metabolic activity: a process known as dormancy, which may last for months, years or even centuries. Dormancy helps keep seeds viable during unfavorable conditions. Upon a return to favorable conditions, seed germination takes place. Favorable conditions could be as diverse as moisture, light, cold, fire, or chemical treatments. After heavy rains, many new seedlings emerge. Forest fires also lead to the emergence of new seedlings. Some seeds require vernalization (cold treatment) before they can germinate. This guarantees that seeds produced by plants in temperate climates will not germinate until the spring. Plants growing in hot climates may have seeds that need a heat treatment in order to germinate, to avoid germination in the hot, dry summers. In many seeds, the presence of a thick seed coat retards the ability to germinate. Scarification, which includes mechanical or chemical processes to soften the seed coat, is often employed before germination. Presoaking in hot water, or passing through an acid environment, such as an animal's digestive tract, may also be employed.

Depending on seed size, the time taken for a seedling to emerge may vary. Species with large seeds have enough food reserves to germinate deep below ground, and still extend their epicotyl all the way to the soil surface. Seeds of small-seeded species usually require light as a germination cue. This ensures the seeds only germinate at or near the soil surface (where the light is greatest). If they were to germinate too far underneath the surface, the developing seedling would not have enough food reserves to reach the sunlight.

Development of Fruit and Fruit Types

After fertilization, the ovary of the flower usually develops into the fruit. Fruits are usually associated with having a sweet taste; however, not all fruits are sweet. Botanically, the term "fruit" is used for a ripened ovary. In most cases, flowers in which fertilization has taken place will develop into fruits, and flowers in which fertilization has not taken place will not. Some fruits develop from the ovary and are known as true fruits, whereas others develop from other parts of the female gametophyte and are known as accessory fruits. The fruit encloses the seeds and the developing embryo, thereby providing it with protection. Fruits are of many types, depending on their origin and texture. The sweet tissue of the blackberry, the red flesh of the tomato, the shell of the peanut, and the hull of corn (the tough, thin part that gets stuck in your teeth when you eat popcorn) are all fruits. As the fruit matures, the seeds also mature.

Fruits may be classified as simple, aggregate, multiple, or accessory, depending on their origin. If the fruit develops from a single carpel or fused carpels of a single ovary, it is known as a simple fruit, as seen in nuts and beans. An aggregate fruit is one that develops from more than one carpel, but all are in the same flower: the mature carpels fuse together to form the entire fruit, as seen in the raspberry. Multiple fruit develops from an inflorescence or a cluster of flowers. An example is the pineapple, where the flowers fuse together to form the fruit. Accessory fruits (sometimes called false fruits) are not derived from the ovary, but from another part of the flower, such as the receptacle (strawberry) or the hypanthium (apples and pears).

Fruits generally have three parts: the exocarp (the outermost skin or covering), the mesocarp (middle part of the fruit), and the endocarp (the inner part of the fruit). Together, all three are known as the pericarp. The mesocarp is usually the fleshy, edible part of the fruit; however, in some fruits, such as the almond, the endocarp is the edible part. In many fruits, two or all three of the layers are fused, and are indistinguishable at maturity. Fruits can be dry or fleshy. Furthermore, fruits can be divided into dehiscent or indehiscent types. Dehiscent fruits, such as peas, readily release their seeds, while indehiscent fruits, like peaches, rely on decay to release their seeds.

There are four main types of fruits. Simple fruits, such as these nuts, are derived from a single ovary. Aggregate fruits, like raspberries, form from many carpels that fuse together. Multiple fruits, such as pineapple, form from a cluster of flowers called an inflorescence. Accessory fruit, like the apple, are formed from a part of the plant other than the ovary.

Fruit and Seed Dispersal

The fruit has a single purpose: seed dispersal. Seeds contained within fruits need to be dispersed far from the mother plant, so they may find favorable and less competitive conditions in which to germinate and grow.

Some fruit have built-in mechanisms so they can disperse by themselves, whereas others require the help of agents like wind, water, and animals. Modifications in seed structure, composition, and size help in dispersal. Wind-dispersed fruit are lightweight and may have wing-like appendages that allow them to be carried by the wind. Some have a parachute-like structure to keep them afloat. Some fruits—for example, the dandelion—have hairy, weightless structures that are suited to dispersal by wind.

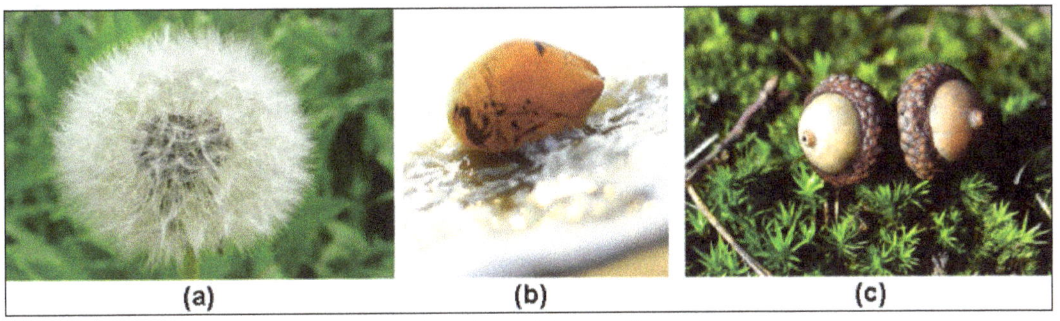

Figure: Fruits and seeds are dispersed by various means.

a) Dandelion seeds are dispersed by wind, the (b) coconut seed is dispersed by water, and the (c) acorn is dispersed by animals that cache and then forget it.

Animals and birds eat fruits, and the seeds that are not digested are excreted in their droppings some distance away. Some animals, like squirrels, bury seed-containing fruits for later use; if the squirrel does not find its stash of fruit, and if conditions are favorable, the seeds germinate. Some fruits, like the cocklebur, have hooks or sticky structures that stick to an animal's coat and are then transported to another place. Humans also play a big role in dispersing seeds when they carry fruits to new places and throw away the inedible part that contains the seeds.

Androecium and Gynoecium are known to be the conceptive structures of a flower. The male conceptive structure is Androecium. It comprises of whorl of stamens.

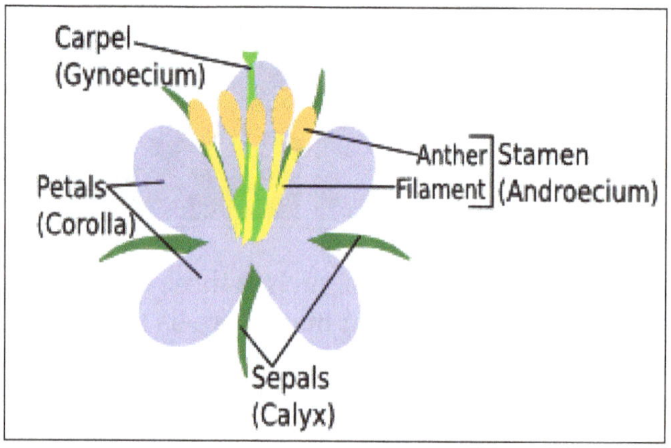

Stamen: It has slim stalk known as Filament. At its tip is contained a bilobed structure called anther. Usually these are bilobed anthers containing two theca (dithecous) isolated by longitudinal groove. The transverse section of any anther shows four sided micro sporangia two in every flap. Microsporangia further form the pollen sac. Contained in the pollen sac are the pollen grains.

Structure of Micro Sporangia

Microsporangium is encompassed by 4 layers:

- Epidermis: External single layer of defensive cells.
- Endothecium: Endothecium is a single celled layer possessing the character of dehiscence. It discharges pollen dust. The character of dehiscence is because of the hygroscopic nature of endothecium.
- Middle layers: It is characterized by three to four layers of cells.
- Tapetum: It is the deepest layer of micro sporangia that provides nourishment to maturing microspores.

Microsporengiogenesis

It is the procedure of development of microspores from dust mother cells through the process of meiotic cell division. The microspores framed are a group of four cells called Microspore Quadruplicate. As there is further development in the anther the moisture from the anther also dries out. The microspores separate and form into dust grains.

Pollen Grains

Pollen Grains or Dust Grains are round in shape. It consists of two layers: external called Exine and internal called Intine. The exine is comprised of a natural compound sporopollenin. It is hard and resistance natural compound.No one protein is known to debase this protein. The exine consists of a typical pore called germ pore.

As the dust grain develops, it contains two cells: a greater Vegetative Cell and minute Generative Cell. The generative cell isolates mitotically into two male gametes.

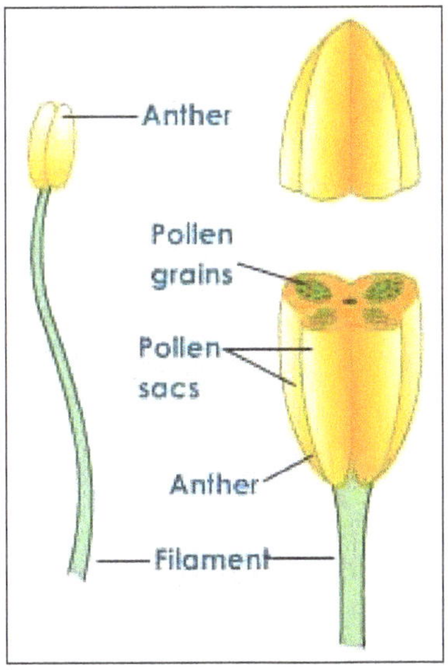

Figure: Structure of Anther.

Gynoecium

The female regenerative part of the flower is the Gynoecium. It comprises of pistil. Gynoecium is termed in many ways:

- Monocarpellary : Gynoecium comprises of single pistil. For Example: Pea, Bean.
- Multicarpellary: Gynoecium comprises of more than one pistil.
    - Syncarpous: Gynoecium with combined pistil. For Example: Tomato, Cucumber.
    - Apocarpous: Gynoecium with free pistil. For Example: Lotus Vinca.

Pistil

Each pistil has three sections - the Stigma, Style and Ovary:

- Stigma: At the tip of pistil is the stigma that provides a landing stage for dust grains.
- Style: The style is the lengthened slender part underneath the stigma.
- Ovary: The basal lump part of the pistil is ovary.

The hole inside the ovary is called Ovarian Hole (Locule). The Megasporangia usually called Ovule emerge from the placenta. The quantity of ovules inside an ovary might be one (mango, paddy, and wheat) or numerous (orchids, watermelon, papaya).

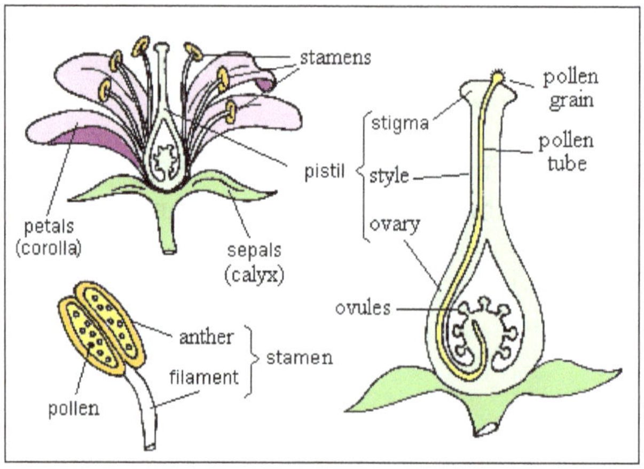

Figure: Gynoecium diagram in a flower.

Figure: Gynoecium diagram.

Megasporangium (Ovule)

The ovule is a minute structure connected to the placenta with a stalk known as Funicle. The ovule's body wires with funicle in the area called Hilum. Each ovule has maybe a couple of defensive envelopes called integuments. Integuments encompass the ovule aside from at the tip. The minute opening at the tip is called Micropyle. Chalaza is a structure present opposite to the micropylar end. It denotes the basal part of the ovule. (explained in the diagram of embryo sac presented).

The integument encloses nucellus which is a mass of cells. These cells have abundant nourishment materials. Inside the nucleus is the embryo sac or female gametophyte. An ovule by and large has a single embryo sac shaped from a megaspore through reduction division.

Megasporogenesis

The procedure of arrangement of megaspores from the megaspore mother cell is known as Megasporogenesis. Ovules for the most part separate a single Megaspore Mother Cell (MMC) in the micropylar area of the nucellus. The MMC experiences meiotic division.

The dominant part of flowering plants, one of the megaspores is utilitarian while the other three deteriorate. This functional megaspore forms into the female gametophyte. The nucellus is by nature diploid. MMC is diploid; the utilitarian haploid, female gametophyte is haploid).

The core of the utilitarian megaspore divides mitotically to shape two nuclei. These two move to the inverse shafts, forming the 2-nucleate embryo sac. Two more successive mitotic atomic divisions result in the arrangement of 8-nucleate phases of the embryo sac.

One nucleus from every shaft comes to centre of embryo sac. They intertwine to frame secondary nucleus. The three nuclei at the chalazal end shapethe antipodal. The three nuclei at the micropylar end sort out in to egg apparatus. They are recognized in to two Synergids with an egg between them. The angiosperm embryo sac, at development, however 8 nucleate but are 7-celled.

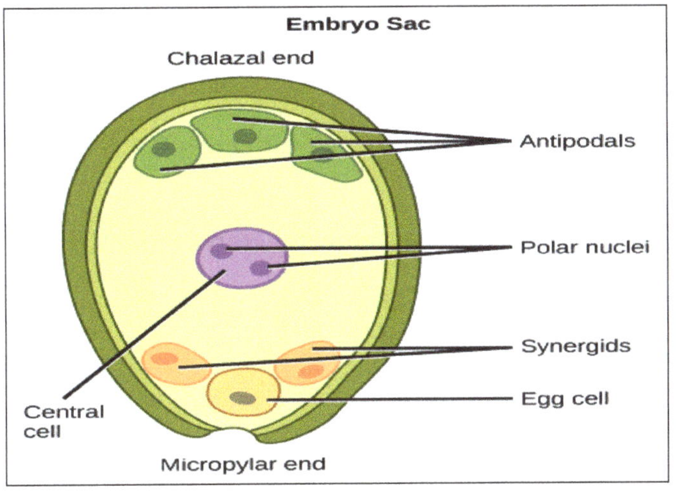

Pollination

Transfer of dust grains to the stigma of a pistil is called Pollination.

Type of Pollination

- Autogamy: Transfer of dust grains from the anther to the stigma of a similar flower is called autogamy. In such flower anthers and the stigma lie near each other with the goal that self-pollination can happen.

Chasmogamous flowers and Cleistogamous flowers: Some plants produce two sorts of flowers. For Example: Viola (Regular Pansy), Oxalis, and Commelina. Flowers with uncovered anthers and stigma are known to be chasmogamous flower. Flowers which don't open are known as Cleistogamous Flowers.

- Geitonogamy: Transfer of dust grains from the anther to the stigma of the diverse flower but of the similar plant is called Geotonogamy. Hereditarily it is like autogamy on the grounds that the dust grains originate from the same plant.
- Xenogamy: Transfer of dust grains from anther to the stigma of an alternate plant but of the same species.

Agents of Pollination

Plants utilize two abiotic (Wind and Water) and one biotic (Animals) operators to accomplish pollination. Majority of plants utilize biotic operators for pollination. Just a little extent of plants utilizes abiotic operators.

- Anemophily: The pollination happened by the impact of wind as an operator is called Anemophily.
- Hydrophily: The pollination happened by the impact of water as an operator in hydrophytes is called Hydrophily.
- Zoophily: The pollination happened by the impact of creatures as an operator is called Zoophily.

On the basis of various sorts of creatures involved in pollination, they are grouped into:

- Entamophily: The pollination happened by the impact of insects as an operator is called Entamophily.
- Ornithophily: The pollination happened by the impact of birds as a specialist is called Ornithophily.
- Chirapterophily: The pollination happened by the impact of bats as a specialist is called Chirapterophily:
    - Anemophily: It is the name given to the pollination by wind. Little flowers are pollinated by anemophily. These plants produce huge measure of dust grains. For Example: Sugar Stick, Grasses, Bamboo, Rice, and so on. The anemophilous flowers are small and never shaded in bright color. They are unscented and never create nectar. Pollen grains are little, dry and light. Stigmas are bushy and spread to trap dust grains floating in air.
    - Hydrophily: It is the pollination by water. Not all aquatic plants experiences hydrophily. When the pollination happens underneath the water level it is called Hypohydrogamic. It found in plants like Hydrilla, Ceratophyllum. When the pollination happens at the water level is called Epihydrogamic. It is found in plants like Vallisneria.

In some other aquatic plants like water hyacinth (Eichhornia), water lily flower develops out of water and pollination happens by wind or insects. In vallisneria, the sub-consolidated aquatic plant, the male flowers opens at the surface of water. The discharged dust grains coasts on water flow. The pedicel of female flower stretches till it achieves the water surface to contact with dust grains.

- Zoophily: It is the pollination by creatures like bats, birds, insects, wasps, honey bees and so on. Majority of plants are pollinated by insects. The entamophilous flowers have some trademark highlights:
    - The flowers are fragrant and overlook scent. For Example: Jasmine, Rose.
    - The flowers are colorful to attracted insects.

- The nectar organ produces nectar to encourage visiting insects.
- The dust grains are sticky or spiny. It attaches to the group of visiting insects.
- To catch dust grains, the stigma has to be sticky.

Yucca Plant is Pollinated by Moth

The moth and yucca can't finish their existence without each other. The moth drills an opening in ovary to lay its eggs. At that point it gathers dust grains from a few flowers and pushes in the honor end of stigma. After fertilization and seed improvement, the hatchlings of moth eat developing seeds. The unconsumed seeds are scattered to spread. Accordingly, moth can't sustain life without moth and yucca can't imitate sexually without moth.

Inbreeding

The fertilization that happens within the single plant is called inbreeding. Larger part of plants is androgynous. The autogamy (self-fertilization) and geitonogamy brings about inbreeding. The significant disadvantage of self-fertilization is, the offspring gets feeble and causes diminished wellness in population. It is called inbreeding wretchedness. To stay away from these, plants create numerous structures that demoralize self-fertilization.

Outbreeding

The fertilization happens between two unique plants of same species is called out breeding. The Xenogamy (cross-pollination) brings about outbreeding. The plants accomplished out-breeding by numerous systems. The Unisexual Flowers are delivered in same plant or distinctive plant. In monoecious plants with unisexual flowers, similar to castor, cucurbita, maize it maintains a strategic distance from autogamy. In dioecious plants like papaya, mulberry, it brings about xenogamy.

Self-Sterility or Incompatibility

- In this dust of the flower has no fertilizing impact on stigma of a similar flower. For Example: Pas-siflora, Potato.
- Dichogamy: In this male and female sex organ develops at various times. Ex: china rose, lady's finger, jasmine, Custard apple.
- Herkogamy: The anther and stigma are put at various positions so that pollens are not able to achieve stigma of same flower.

Artificial Hybridization

Pollinating the dust grains of one flower to the next flower of various plants of same species is known as Artificial Hybridization. It is one of the significant approaches of harvest improvement program. In this strategy, the required dust grains are utilized for pollination. The stigma is shielded from contamination with other dust grains. It is accomplished by castration and bagging.

Emasculation: It is the expulsion of anthers/stamens from flower bud before it develops.

Figure: Emasculation diagram.

Bagging: It is the covering of undermined flower with margarine paper to avert contamination with different pollens.

Figure: Different methods of Bagging.

Pollen Pistil Interaction: It is the capacity of pistil to perceive the pollen of same species. It experience post pollination occasion that prompts to fertilization. If wrong pollen grain (different species) is arrived on stigma, pistil rejects by preventing dust germination. The acknowledgment and dismissal is interceded by the compound part of dust grain interacting with pistil.

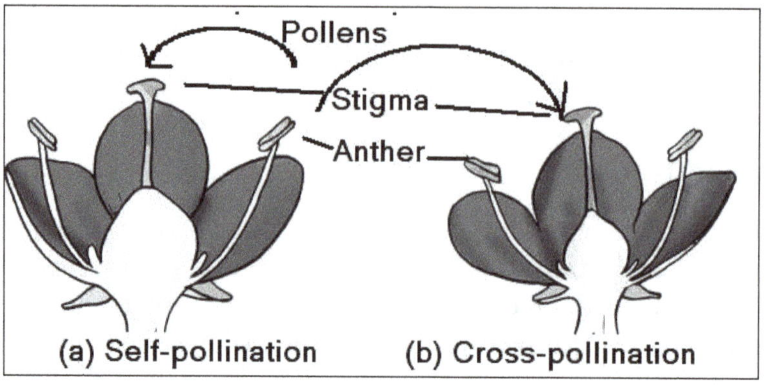

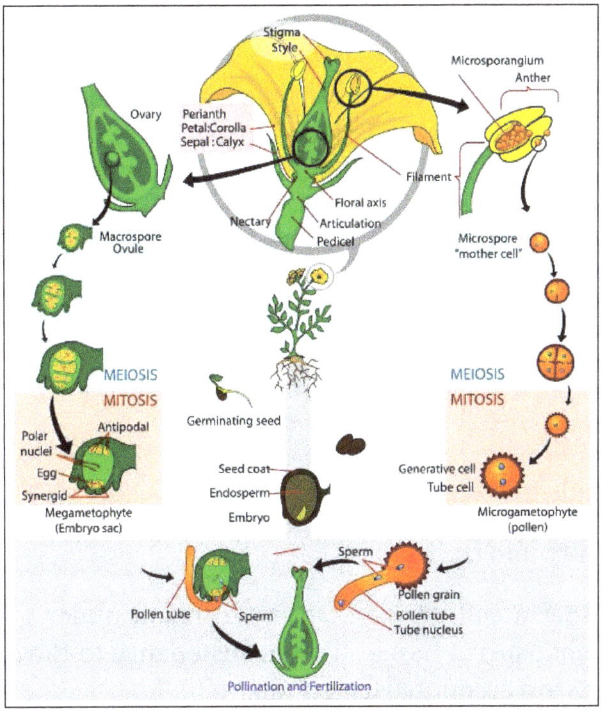

Figure: Pollination and Fertilization.

Fertilization

The fertilization in angiosperms is named as Double Fertilization and Triple Fusion. As the pollen-grain develops as a rule it contains One Vegetative Cell and Two Generative Cells that form into male gamete. In a few plants dust grain sheds as two cell condition. In this, generative cell partitions into two male gametes as pollen tube shapes.

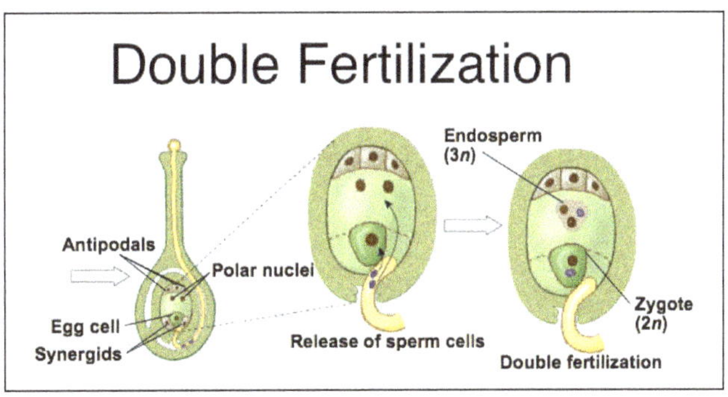

As the dust grain arrives on stigma, it begins to germinate. The pollen tube create leaves germ pore. It develops through the tissue of stigma, style and infiltrates the ovule. The section of pollen tube in to ovule is from a distinct place and is distinguished into:

- Porogamy: It is the passage of dust tube through micropyle.
- Chalazogamy: It is the passage of dust tube through chalaza.
- Misogamy: It is the section of dust tube through integument or funicle.

As pollen tube enters the ovule, the tip breaks inside the embryo sac leading to the release of two male gametes. One of the male gamete wires with egg and structures diploid zygote. Another male gamete wires with diploid secondary nucleus to frame triploid endosperm. The fusion of two polar nuclei (to frame secondary diploid nucleus) and male gamete is named as Triple Fusion. The two fusions as zygote development and triple fusion is named as Double Fertilization.

Post-fertilization Stages

It is the series of occasions that happens after double fertilization. It includes:

- Endosperm development.
- Embryo improvement.
- Development of ovule into seed.
- Development of ovary into fruit.

1. Endosperm: The endosperm cell, shaped by triple fusion under goes rehashed mitotic cell division to frame triploid endosperm tissue. It gives sustenance to the developing embryo. Three unique sorts of endosperms are accounted as follows:

- Nuclear endosperm: In this endosperm cell nucleus experiences free nuclear division. Every one of these nuclei is pushed to the fringe forming central vacuole. The cytokines is brings about the development of endosperm tissue. Example: coconut. The vacuole is loaded with liquid.
- Cellular endosperm: In this endosperm cellular division experiences cytokinesis. Example: Datura.
- Helobial endosperm: In first endosperm cell division two unequal cells shapes. One cell experiences cellular division and other undergo free nuclear division. Example: monocots. In pea, groundnut, beans endosperm is totally devoured by developing embryo. In a few seeds like groundnut, castor endosperm preserves developed seeds.

2. Embryo improvement.

- In Dicot: The zygote separates transversely in to upper suspensor cell exhibited at micropyle area and lower embryonal cell exhibited at chalazal district. The suspensor cell partitions and structures 7 – 10 cells called suspensor. The upper cell of suspensor extends and appends to the embryo sac. The embryonal cell separates into eight cells. The lower four cells creates in to form one plumule and two cotyledons. The above four cells close to the suspensor creates in to form hypocotyl and radicle. The completely created embryo of dicotyledons has an embryonal hub that separated into plumule, two cotyledons and a radicle. The lower end of radicle (root tip) is secured with root cap. In the starting embryo is globular in shape. At the point when two cotyledons separate it frames heart shape.
- In monocots: The zygote isolates transversely in to upper suspensor cell display at micropyle locale and lower embryonal cell exhibited at chalazal district. The embryonal

cell partitions into globular embryo. It frames a huge cotyledon and plumule. Plumule is covered by coleoptile. The single cot-yledon is called scutellum. At lower end of embryonal pivot radicle creates. It is secured by a sheath called coleorhiza.

3. Seed: (refer to pollination and fertilization diagram above) The final result of sexual reproduction is the formation of seed. It is the ovule that is fertilized. Seed comprises of seed coat, cotyledon and embryonal hub. The cotyledons are swollen because of their capacity of hold sustenance. The seeds are either albuminous or non-albuminous. In albuminous seed endosperm is not totally utilized by embryo. Example: wheat, maize, castor, sunflower. In a few seeds like dark pepper the nucellus is additionally present. The leftover persistent nucellus is called Perisperm. In non albuminous seed endosperm is totally utilized by embryo. Example: Pea, Groundnut. The seed coat is developed from the integument of ovule. The micro-pyle remains as a pore in the seed coat. It encourages the section of oxygen and water in to seed during germination. Seeds become scarce by losing water. The embryo enters dormancy, the phase of inactivity. As the conditions become favorable again it begins germination.

4. Fruit: The wall of ovary forms into organic product wall called pericarp. It encompasses the seed. In apple, strawberry the fruit is developed from the thalamus. Subsequently these are termed false fruit. The fruits created from unfertilized ovary are called Parthenocarpic Fruits. Parthenocarpic fruits are seedless.

Apomixes

Apomixes is the process of formation of seed without the fertilization of ovule and the fruits created from unfertilized ovule are termed apomictic fruits. The descendants of hybrid isolate their characters and not continue the hybrid character till a great many generations. Consequently the ranchers ought to purchase hybrid seeds each year. It makes broad. The apomixes is performed in hybrid plants to maintain a strategic distance from the isolation of characters. The farmers can utilize hybrid seeds to raise new yield each year.

Polyembryony: The nearness of more than one embryo in seed is known as Polyembryony. Example: Seeds of citrus fruits.

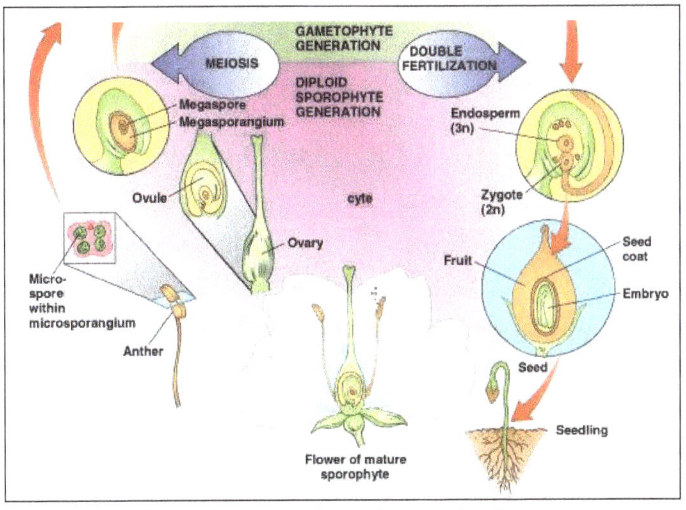

Figure: Process of Fertilization in plants.

References

- Plant-reproduction: letstalkscience.ca, Retrieved 05, May 2020
- Asexual-Reproduction, plant-reproduction, general-biology-(OpenStax), Introductory-and-General-Biology: bio.libretexts.org, Retrieved 21, January 2020
- Apomixis-Developmental-Characteristics-and-Genetics-253093824: researchgate.net, Retrieved 07, August 2020
- Sexual-reproduction-plants, biology: byjus.com, Retrieved 19, March 2020
- Sexual-reproduction-in-plants: courses.lumenlearning.com, Retrieved 04, January 2020
- Sexual-reproduction-in-flowering-plants: askiitians.com, Retrieved 27, June 2020

Chapter 3

Molecular Breeding in Plants

Plant molecular breeding may be broadly defined as the use of genetic manipulation at DNA molecular levels in order to alter or improve certain desirable characteristics in plants. Molecular markers are specific DNA fragments in a genome which are used in marker-assisted breeding to achieve desired changes. All aspects of molecular breeding in plants are discussed in detail in this chapter.

Plant Molecular Breeding

Plant breeding is simple: cross the best parents, and identify and recover progeny that outperform the parents. In practice, plant breeding is a three step process, wherein populations or germplasm collections with useful genetic variation are created or assembled, individuals with superior phenotypes are identified, and improved cultivars are developed from selected individuals. A wide diversity of approaches, tailored to the crop species and breeding objectives, have been developed for improving cultivars. These breeding methods feature different types of populations, selection procedures, and outcomes.

Figure summarizes the three breeding methods that are commonly employed in crop improvement programs. When the goal is to upgrade an established elite genotype with trait(s) controlled by one or a few loci, backcrossing is used either to introgress a single gene or to pyramid a few genes. For genetically complex traits, germplasm improvement instead requires reshuffling of the genome to produce new favorable gene combinations in the progeny. The pedigree breeding method produces such novelty via crossing and recombination among superior, yet complementary, parents and selection among segregating progeny for improved performance. Recurrent selection aims to simultaneously increase the frequencies of favorable alleles at multiple loci in breeding populations through intermating of selected individuals. For hybridized crops such as maize, recurrent selection may be extended to improve the performance of distinct complementary populations (e.g. heterotic groups) that are used as parents to form superior hybrid combinations. This practice is referred to as reciprocal recurrent selection.

Quantitative genetic principles have been particularly powerful as the theoretical basis for both population improvement and methods of selecting and stabilizing desirable genotypes. An important concept in quantitative genetics and plant breeding is genetic gain (ΔG), which is the predicted change in the mean value of a trait within a population that occurs with selection. Regardless of species, the trait of interest, or the breeding methods employed, ΔG serves as a simple universal expression for expected genetic improvement. Figure below shows the genetic gain equation and an expansion of its terms to fundamental parameters of quantitative genetics. Though clearly an oversimplification of the advanced quantitative genetic principles employed in plant breeding, the genetic gain equation effectively relates the four core factors that influence breeding progress: the degree of phenotypic variation present in the population (represented by its sd, σ_p), the probability that a trait phenotype will be transmitted from parent to offspring (heritability, h^2), the proportion of the population selected as parents for the next generation (selection intensity, i, expressed in units of sd from the mean), and the length of time necessary to complete a cycle of selection (L). L is not only a function of how many generations are required to complete a selection cycle, but also how quickly the generations can be completed and how many generations can be completed per year.

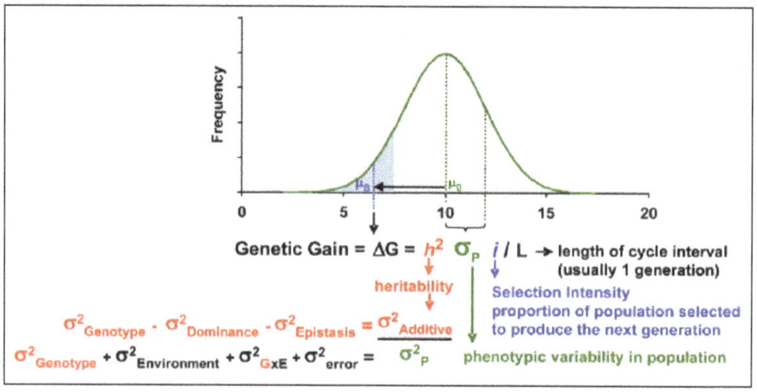

Figure: The genetic gain equation and its component variables.

The top portion illustrates an idealized distribution showing the frequency of individuals within a breeding population (y axis) that exhibit various classes of phenotypic values (x axis). Mean phenotypic value (μo) of the original population (shown as entire area under the normal curve) and mean (μS) for the group of selected individuals (shaded in blue) are indicated. In this generalized example, trait improvement is achieved by selecting for a lower phenotypic value, e.g. grain moisture at harvest in maize. Components of variation (σ2) that contribute to the sd of the phenotypic distribution (σP) are indicated below the histogram.

It is clear that ΔG can be enhanced by increasing σ_p, h^2, or i, and by decreasing L. Thus, the genetic gain equation provides a framework for comparing the predicted effectiveness of particular breeding strategies and is often used as a guide to the judicious allocation of resources for achieving breeding objectives. When considered in the context of the genetic gain concept, molecular plant breeding offers powerful new approaches to overcome previous limitations in maximizing ΔG. The following sections cite examples where molecular plant breeding positively impacts ΔG and each of its component variables. For brevity, we focus on examples from maize where molecular breeding is most advanced, and has now become the primary means to develop improved commercial hybrids.

Molecular Plant Breeding Expands Useful Genetic Diversity for Crop Improvement

The maximum potential for genetic gain is proportional to the phenotypic variation (σ_p) present in the original source population and maintained in subsequent cycles of selection. Phenotypic variation is positively associated with genetic diversity, yet also depends on environmental factors and the interactions between genotype and environment. Genetic diversity may be derived from breeding populations (either naturally occurring or synthetic), segregating progeny from a cross of selected parental lines, exotic materials that are not adapted to the target environment, wide interspecific crosses, naturally occurring or induced mutations, the introduction of transgenic events, or combinations of these sources.

However, not all phenotypic variation is equal. For example, the use of exotic germplasm has been extremely successful for improving many crop species, but difficulties may be encountered through the introduction of undesirable alleles associated with lack of adaptation. The need for genetic diversity must be balanced by elite performance, because choosing the best parents is key to maximizing the probability for successful improvement. In contrast, the expected increase in linkage disequilibrium among elite populations derived from intense prior selection may also limit the creation of new genetic combinations for future gain. Intermating source populations for genetic recombination may overcome this problem, but delays cultivar development.

Molecular markers and more recently, high-throughput genome sequencing efforts, have dramatically increased knowledge of and ability to characterize genetic diversity in the germplasm pool for essentially any crop species. Using maize as one example, surveys of molecular marker alleles and nucleotide sequence variation have provided basic information about genetic diversity before and after domestication from its wild ancestor teosinte, among geographically distributed landraces, and within historically elite germplasm.

While molecular markers and other genomic applications have been highly successful in characterizing existing genetic variation within species, plant biotechnology generates new genetic diversity that often extends beyond species boundaries. Biotechnology enables access to genes heretofore not available through crossing and creates an essentially infinite pool of novel genetic variation. Genes may be acquired from existing genomes spanning all kingdoms of life, or designed and assembled de novo in the laboratory. Both subtle and extreme examples of the power of transgenes to introduce novel phenotypic variation can be found in the three different transgenes developed for resistance to glyphosate herbicides in maize and other crops. The first glyphosate-tolerant maize hybrids used a modified version of the endogenous maize gene encoding 5-enol-pyruvylshikimate-3-P synthase, which was followed later by events produced with a 5-enol-pyruvylshikimate-3-P synthase gene isolated from Agrobacterium. More recently, a synthetic gene with enhanced glyphosate acetyltransferase activity was created via gene shuffling and selection in a microbial system. Each of these glyphosate-tolerant maize events also illustrates another benefit of biotechnology, where new combinations of regulatory sequences (e.g. the cauliflower mosaic virus 35S and rice actin1 promoters) may be used to achieve optimal trait expression with respect to overall activity and tissue distribution relative to what might be possible with endogenous genes.

Molecular Plant Breeding Increases Favorable Gene Action

Quantitative genetics uses the theoretical concept of heritability to quantify the proportion of phenotypic variation that is controlled by genotype. In practice, heritability is greatly influenced by the genetic architecture of the trait of interest, which is described by the number of genes, the magnitude of their effects, and the type of gene action associated with phenotypes. Better knowledge of genetic architecture and favorable gene action (that which is more amenable to selection) often has the greatest impact on improving genetic gain. For the genetic gain formula, heritability (h^2) is used in its narrow sense, representing the proportion of phenotypic variation due to additive genetic effects (those that reflect changes in allele dosage or allelic substitutions). Additive genetic effects are also referred to as the breeding value because they are predictably transmitted to progeny. Deviations from additive effects are significant for many traits, and are partitioned into either dominance effects that reflect the interactions between different alleles at the same locus or epistatic effects resulting from interactions among different loci. Gene action and breeding values are characterized by progeny testing, where the phenotypes of individuals in a population are compared to their parents and siblings produced from either self-pollination or outcrossing.

Previous efforts to develop large numbers of molecular markers, high density genetic maps, and appropriately structured mapping populations have now made routine for many crop species the ability to simultaneously define gene action and breeding value at hundreds and often thousands of loci distributed relatively uniformly across entire genomes. The results from such mapping studies provide greatly improved estimates for the number of loci, allelic effects, and gene action controlling traits of interest. More importantly, genomic segments can be readily identified that show statistically significant associations with quantitative traits (quantitative trait loci [QTLs]). In addition to genetic mapping in families derived from biparental crosses, new advances in association genetics with candidate genes and approaches that combine linkage disequilibrium analysis in families and populations further enhance power for QTL discovery.

Information about QTLs can be used in a number of ways to increase heritability and favorable gene action. For traits exhibiting low to moderate heritability, such as grain yield, QTLs, and their associated molecular markers often account for a greater proportion of the additive genetic effects than the phenotype alone. Furthermore, knowledge of genetic architecture can be exploited to add or delete specific alleles that contribute to breeding value. When either genetic linkage or epistasis among loci with antagonistic effects on a trait limits genetic gain, QTL information can be used to break these undesirable allelic relationships.

Success in using information about QTLs to increase genetic gain depends greatly on the magnitude of QTL effects, precise estimation of QTL positions, stability of QTL effects across multiple environments, and whether QTLs are robust across relevant breeding germplasm. Prediction of QTL positions is enhanced by further fine mapping, which facilitates testing QTL effects and breeding values in additional populations. When the density of observed recombinations approaches the resolution of single genes, the causal genetic change for a QTL can be determined. Molecular isolation of QTLs permits the development of perfect or functional molecular markers at the potential resolution of the fundamental unit of inheritance, the nucleotide, and dramatically increases the specificity and precision by which genetic effects are estimated and manipulated in breeding programs.

The use of transgenes can further simplify the genetic architecture for desirable traits, in ways that may be superior to or not possible even when perfect markers are available for robust QTLs of large effect. Transgenes typically condition strong genetic effects at operationally single loci, which also exhibit dominant gene action where only one copy of the event is needed for maximal trait expression in a hybrid cultivar. These features of transgenes can reduce complex quantitative improvement to a straightforward, often dramatic, solution. Excellent examples are provided by the expression in transgenic corn hybrids of insecticidal toxin proteins from Bacillus thuringiensis (Bt) to reduce feeding damage by larvae of the European corn borer (Ostrinia nubilalis) or the corn rootworm beetle (Diabrotica spp.). Partial resistance in maize germplasm to these insect pests had been previously characterized as quantitatively inherited traits with low heritability, but the Bt transgenic events offer a simply inherited alternative that is efficiently manipulated in breeding programs.

By simplifying genetic architecture, transgenes may also permit disruption of allelic interactions between factors controlling the trait of interest and other important performance characteristics. For example, employing a transgenic source of insect resistance (e.g. a single locus Bt transgene) may facilitate selection for favorable alleles for yield improvement that are tightly linked in repulsion with endogenous genes for resistance to the same class of insect pests. In addition, transgenic events may be engineered to uncouple negative pleiotropic effects from beneficial phenotypes conditioned by recessive mutations. This application is illustrated by the use of RNA interference to specifically down-regulate zein seed storage protein gene expression. This strategy mimics the effects of the opaque2 mutation on improving the amino acid profile of maize grain for animal feed, while circumventing the softer endosperm texture and susceptibility to fungal pathogens typically associated with opaque2.

Transgenic events can also be designed to intervene at key regulatory steps for entire metabolic or developmental pathways, such that gene action for the corresponding traits are largely inherited as single dominant factors that are less sensitive to environmental effects. Examples include the expression of a transcription factor that increases drought tolerance, and altering the balance between levels of the GLOSSY15 transcription factor relative to its repressor, microRNA172, to delay flowering time in maize hybrids.

Biotechnology also facilitates the molecular stacking of transgenes that control a trait or suite of traits into a single locus haplotype defined by a transgenic event. Examples of such an approach include the initial Golden Rice, recently released YieldGuard VT Triple transgenic maize hybrids where herbicide tolerance and multiple insect resistance traits are integrated as one genomic locus, or the combination of transgenes that simultaneously increase synthesis and decrease catabolism of Lys in maize seeds. Recent reports of improvements in gene targeting technology and the construction of meiotically transmissible plant minichromosomes pave the way for introducing more traits with increasing complexity. With such advances, biotechnology is now poised to assemble useful genetic diversity from essentially any source into constructs that concentrate favorable gene action and maximize heritability for a greatly expanded set of traits.

The molecular cloning of QTLs has yielded novel insights about the biology of quantitative traits that were not likely to be discovered from the analysis of gene knockouts or overexpression strategies, in particular the impacts of regulatory variation on phenotypic variation and evolution . Furthermore, molecular markers, genomics, and biotechnology are now applied in an iterative network to exploit genetic diversity for crop improvement. Genomic information enables the discovery of beneficial alleles via QTL mapping and cloning, followed by the use of information learned from the molecular characterization of QTLs to design optimal transgenic strategies for crop improvement.

Molecular Plant Breeding Increases the Efficiency of Selection

Conventional plant breeding that relies only on phenotypic selection has been historically effective. However, for some traits, phenotypic selection has made little progress due to challenges in measuring phenotypes or identifying individuals with the highest breeding value. The effects of environment, genotype by environment interaction, and measurement errors also contribute to observed differences. Evaluation of genotypes in multiple environments with replicated designs allows better estimation of breeding values, but requires additional time and expense. For some traits, it may be necessary to sacrifice the individual to measure phenotypes, or trait expression may depend on variable environmental conditions (e.g. disease pressure) and the stage of development (e.g. grain quality can only be assessed after flowering). Furthermore, plant breeders typically must simultaneously improve a suite of commercially valuable traits, which may limit gains from selection. Just as molecular plant breeding helps to expand genetic diversity, characterize genetic architecture, and modify gene action, its methods can also be applied to increasing the efficiency of selection.

Molecular marker genotypes that are either within genes or tightly linked to QTL influencing traits under selection can be employed as a supplement to phenotypic observations in a selection index. In cases where genetic correlations are high, further efficiencies can be gained by substituting genotypic for phenotypic selection during some selection cycles, which can reduce phenotyping efforts and cycle times by permitting the use of off-season nurseries. Johnson summarized an early example of combining phenotypic data and molecular marker scores to increase selection gains for maize grain yield and resistance to European corn borer. An effective strategy to simultaneously modify multiple traits is the use of selection indices that consider multiple factors in choosing the final improved genotype. Eathington et al. recently reported on results obtained from the use of multiple trait indices and marker-assisted selection for nearly 250 unique corn breeding populations. Use of molecular markers increased breeding efficiency approximately 2-fold relative to phenotypic selection alone, with similar gains also observed in soybean (Glycine max) and sunflower (Helianthus annuus) populations.

Marker-assisted selection can also significantly enhance genetic gain for traits where the phenotype is difficult to evaluate because of its expense or its dependence on specific environmental conditions. Molecular markers may be used to increase the probability of identifying truly superior genotypes, by focusing testing resources on genotypes with the greatest potential (i.e. early elimination of inferior genotypes), by decreasing the number of progeny that must be screened to recover a given level of gain, and by enabling simultaneous improvement for traits that are negatively correlated. Successful examples include resistance to soybean cyst nematode resistance to cereal diseases, and drought tolerance in maize.

The efficiency of phenotypic selection for some complex traits can be enhanced by including physiological or biochemical phenotypes as secondary traits, if these exhibit strong genetic correlations with the target trait and possess high heritability. Recent advances in functional genomics permit the population-scale profiling of RNA abundance, protein levels and activities, and metabolites that are associated with important traits. In addition to molecular markers that tag DNA sequence variation, such genetical genomics approaches may provide additional secondary phenotypes as selection targets, particularly for traits defined by responses to environmental, developmental, or physiological cues.

Marker-assisted selection also accelerates the deployment of transgenes in commercial cultivars. Typically, this has been achieved through marker-assisted backcrossing. However, for future biotechnology improvements such as tolerance to drought or nutrient limitation, forward breeding may be required to co-optimize transgene expression and genetic background because endogenous genes and environmental factors may have the potential to influence the phenotypes resulting from transgenic modifications. Of course, use of molecular markers could aid forward breeding efforts as well. Alternatively, discovery efforts for additional genes or QTLs that are necessary for dependable trait performance may suggest design of new transgene constructs that stack primary transgenes with known genetic modifiers into second-generation transgenic events.

Increasing Adoption of Molecular Plant Breeding

The adoption of molecular plant breeding approaches has occurred at different rates among crop species and institutions engaged in crop improvement, due to the combined influence of scientific, economic, and sociological factors. Important early scientific barriers included the recalcitrance of cereal crop species to Agrobacterium-mediated transformation and lack of knowledge about genetic control of traits already defined as important breeding targets. Continued research and technology development has largely overcome obstacles for plant transformation of nearly all important crop and horticultural species. Similarly, information gained from genomics research in plant species and other organisms has generated a wealth of information about gene structure and function, as well as large numbers of molecular markers for use in plant breeding. Despite these resources, genetic specificity of robust QTLs remains elusive, unless breeding programs and associated information management systems are restructured to fully integrate knowledge of pedigrees, phenotypes, and marker genotypes that can be leveraged to optimize response to selection. Even with such integration, modifying regulatory functions remains a scientific challenge for molecular breeding because it is difficult to determine the sequence basis for regulatory changes and to predict their phenotypic effects.

Once enabling technologies in biotechnology and genomics become available, economic factors often dictate the degree to which these innovations are integrated into existing plant breeding programs. The expense of gaining governmental regulatory approval for commercial release of transgenic varieties is a significant economic barrier. The costs associated with the development, establishment, and operation of molecular plant breeding are greater than conventional plant breeding practices, requiring significant investments in new research infrastructure and intellectual capacity. Such resources initially existed only in private agribusiness firms and a handful of larger public institutions, further accelerating an ongoing trend for increased industrialization of plant breeding programs among major crops such as corn, soybeans, cotton (Gossypium hirsutum), and wheat. Where there has been adoption in companies, the balance may favor biotechnology over QTLs for improving complex traits, despite greater product development costs, because transgenics can be designed to produce stronger, even dramatic, phenotypic effects, and can often be more rapidly deployed across a broader range of germplasm, resulting in new solutions sooner.

Though molecular breeding is now considered an essential component of current crop improvement efforts for major crops by large companies, the broad applicability of modern molecular approaches to conventional plant breeding remains a source of debate among some practicing plant breeders in the public sector, particularly for minor crops. In addition to the valid scientific and economic factors that have delayed or prevented adoption of molecular approaches in meeting some plant breeding objectives, there are at least three additional reasons contributing to this view. First, molecular plant breeding requires training and expertise in both molecular biology and plant breeding. Educational efforts that delivered such interdisciplinary training were initially established in the early 1990s, but still remain limited to a relatively small group of academic institutions with historic strengths in plant breeding. A second reason for reduced enthusiasm to embrace biotechnology among some plant breeders is the problems with acceptance of transgenic crops among certain governments and groups of consumers, as exemplified by the shelving of wheat varieties with transgenes for resistance to glyphosate herbicides.

Molecular Marker and Marker Assisted Breeding

Molecular breeding (MB) may be defined in a broad-sense as the use of genetic manipulation performed at DNA molecular levels to improve characters of interest in plants and animals, including genetic engineering or gene manipulation, molecular marker-assisted selection, genomic selection, etc. More often, however, molecular breeding implies molecular marker-assisted breeding (MAB) and is defined as the application of molecular biotechnologies, specifically molecular markers, in combination with linkage maps and genomics, to alter and improve plant or animal traits on the basis of genotypic assays. This term is used to describe several modern breeding strategies, including marker-assisted selection (MAS), marker-assisted backcrossing (MABC), marker-assisted recurrent selection (MARS), and genome-wide selection (GWS) or genomic selection (GS).

Genetic Markers in Plant Breeding: Conceptions, Types and Application

Genetic markers are the biological features that are determined by allelic forms of genes or genetic loci and can be transmitted from one generation to another, and thus they can be used as experimental probes or tags to keep track of an individual, a tissue, a cell, a nucleus, a chromosome or a gene. Genetic markers used in genetics and plant breeding can be classified into two categories: classical markers and DNA markers. Classical markers include morphological markers, cytological markers and biochemical markers. DNA markers have developed into many systems based on different polymorphism-detecting techniques or methods (southern blotting – nuclear acid hybridization, PCR – polymerase chain reaction, and DNA sequencing), such as RFLP, AFLP, RAPD, SSR, SNP, etc.

Classical Markers

Morphological markers: Use of markers as an assisting tool to select the plants with desired traits had started in breeding long time ago. During the early history of plant breeding, the markers used mainly included visible traits, such as leaf shape, flower color, pubescence color, pod color, seed color, seed shape, hilum color, awn type and length, fruit shape, rind (exocarp) color and stripe, flesh color, stem length, etc. These morphological markers generally represent genetic polymorphisms which are easily identified and manipulated. Therefore, they are usually used in construction of linkage maps by classical two- and/or three-point tests. Some of these markers are linked with other agronomic traits and thus can be used as indirect selection criteria in practical breeding. In the green revolution, selection of semi-dwarfism in rice and wheat was one of the critical factors that contributed to the success of high-yielding cultivars. This could be considered as an example for successful use of morphological markers to modern breeding. In wheat breeding, the dwarfism governed by gene Rht10 was introgressed into Taigu nuclear male-sterile wheat by backcrossing, and a tight linkage was generated between Rht10 and the male-sterility gene Ta1. Then the dwarfism was used as the marker for identification and selection of the male-sterile plants in breeding populations . This is particularly helpful for implementation of recurrent selection in wheat. However, morphological markers available are limited, and many of these markers are not associated with important economic traits (e.g. yield and quality) and even have undesirable effects on the development and growth of plants.

Cytological markers: In cytology, the structural features of chromosomes can be shown by chromosome karyotype and bands. The banding patterns, displayed in color, width, order and position, reveal the difference in distributions of euchromatin and heterochromatin. For instance, Q bands are produced by quinacrine hydrochloride, G bands are produced by Giemsa stain, and R bands are the reversed G bands. These chromosome landmarks are used not only for characterization of normal chromosomes and detection of chromosome mutation, but also widely used in physical mapping and linkage group identification. The physical maps based on morphological and cytological markers lay a foundation for genetic linkage mapping with the aid of molecular techniques. However, direct use of cytological markers has been very limited in genetic mapping and plant breeding.

Biochemical/Protein Markers

Protein markers may also be categorized into molecular markers though the latter are more referred to DNA markers. Isozymes are alternative forms or structural variants of an enzyme that have different molecular weights and electrophoretic mobility but have the same catalytic activity or function. Isozymes reflect the products of different alleles rather than different genes because the difference in electrophoretic mobility is caused by point mutation as a result of amino acid substitution. Therefore, isozyme markers can be genetically mapped onto chromosomes and then used as genetic markers to map other genes. They are also used in seed purity test and occasionally in plant breeding. There are only a small number of isozymes in most crop species and some of them can be identified only with a specific strain. Therefore, the use of enzyme markers is limited.

Another example of biochemical markers used in plant breeding is high molecular weight glutenin subunit (HMW-GS) in wheat. Payne et al discovered a correlation between the presence of certain HMW-GS and gluten strength, measured by the SDS-sedimentation volume test. On this basis, they designed a numeric scale to evaluate bread-making quality as a function of the described subunits (Glu-1 quality score). Assuming the effect of the alleles to be additive, the Bread-making quality was predicted by adding the scores of the alleles present in the particular line. It was established that the allelic variation at the Glu-D1 locus have a greater influence on bread-making quality than the variation at the others Glu-1 loci. Subunit combination 5+10 for locus Glu-D1 (Glu-D1 5+10) renders stronger dough than Glu-D1 2+12, largely due to the presence of an extra cysteine residue in the Dx-5 subunit compared to the Dx-2 subunit, which would promote the formation of polymers with larger size distribution. Therefore, breeders may enhance the bread-making quality in wheat by selecting subunit combination Glu-D1 5+10 instead of Glu-D1 2+12. Of course, the variation of bread-making quality among different varieties cannot be explained only by the variation in HMW-GS composition, because the low molecular weight glutinen subunit (LMW-GS) (as well as the gliadins in a smaller proportion) and their interactions with the HMW-GS also play an important role in the gluten strength and bread-making quality.

DNA Markers

DNA markers are defined as a fragment of DNA revealing mutations/variations, which can be used to detect polymorphism between different genotypes or alleles of a gene for a particular sequence of DNA in a population or gene pool. Such fragments are associated with a certain location within the genome and may be detected by means of certain molecular technology. Simply speaking, DNA marker is a small region of DNA sequence showing polymorphism (base deletion, insertion and substitution) between different individuals. There are two basic methods to detect the polymorphism: Southern blotting, a nuclear acid hybridization technique, and PCR, a polymerase chain reaction technique. Using PCR and/or molecular hybridization followed by electrophoresis (e.g. PAGE – polyacrylamide gel electrophoresis, AGE – agarose gel electrophoresis, CE – capillary electrophoresis), the variation in DNA samples or polymorphism for a specific region of DNA sequence can be identified based on the product features, such as band size and mobility. In addition to Sothern blotting and PCR, more detection systems have been also developed. For instance, several new array chip techniques use DNA hybridization combined with labeled nucleotides, and new sequencing techniques detect polymorphism by sequencing. DNA markers are also called molecular markers in many cases and play a major role in molecular breeding. Therefore, molecular markers in this article are mainly referred to as DNA markers except specific definitions are given, although isozymes and protein markers are also molecular markers. Depending on application and species involved, ideal DNA markers for efficient use in marker-assisted breeding should meet the following criteria:

- High level of polymorphism.
- Even distribution across the whole genome (not clustered in certain regions).
- Co-dominance in expression (so that heterozygotes can be distinguished from homozygotes).
- Clear distinct allelic features (so that the different alleles can be easily identified).
- Single copy and no pleiotropic effect.
- Low cost to use (or cost-efficient marker development and genotyping).
- Easy assay/detection and automation.
- High availability (un-restricted use) and suitability to be duplicated/multiplexed (so that the data can be accumulated and shared between laboratories).
- Genome-specific in nature (especially with polyploids).
- No detrimental effect on phenotype.

Since first used DNA restriction fragment length polymorphism (RFLP) in human linkage mapping, substantial progress has been made in development and improvement of molecular techniques that help to easily find markers of interest on a large-scale, resulting in extensive and successful uses of DNA markers in human genetics, animal genetics and breeding, plant genetics and breeding, and germplasm characterization and management. Among the techniques that have been extensively used and are particularly promising for application to plant breeding, are the restriction fragment length polymorphism (RFLP), amplified fragment length polymorphism (AFLP), random amplified polymorphic DNA (RAPD), microsatellites or simple sequence repeat

(SSR), and single nucleotide polymorphism (SNP). According to a causal similarity of SNPs with some of these marker systems and fundamental difference with several other marker systems, the molecular markers can also be classified into SNPs (due to sequence variation, e.g. RFLP) and non-SNPs (due to length variation, e.g. SSR). The marker techniques help in selection of multiple desired characters simultaneously using F_2 and back-cross populations, near isogenic lines, doubled haploids and recombinant inbred lines.

RFLP Markers

RFLP markers are the first generation of DNA markers and one of the important tools for plant genome mapping. They are a type of Southern-Boltting-based markers. In living organisms, mutation events (deletion and insertion) may occur at restriction sites or between adjacent restriction sites in the genome. Gain or loss of restriction sites resulting from base pair changes and insertions or deletions at restriction sites within the restriction fragments may cause differences in size of restriction fragments. These variations may cause alternation or elimination of the recognition sites for restriction enzymes. As a consequence, when homologous chromosomes are subjected to restriction enzyme digestion, different restriction products are produced and can be detected by electrophoresis and DNA probing techniques.

RFLP markers are powerful tools for comparative and synteny mapping. Most RFLP markers are co-dominant and locus-specific. RFLP genotyping is highly reproducible, and the methodology is simple and no special equipment is required. By using an improved RFLP technique, i.e., cleaved amplified polymorphism sequence (CAPS), also known as PCR-RFLP, high-throughput markers can be developed from RFLP probe sequences. Very few CAPS are developed from probe sequences, which are complex to interpret. Most CAPS are developed from SNPs found in other sequences followed by PCR and detection of restriction sites. CAPS technique consists of digesting a PCR-amplified fragment and detecting the polymorphism by the presence/absence of restriction sites. Another advantage of RFLP is that the sequence used as a probe need not be known. All that a researcher needs is a genomic clone that can be used to detect the polymorphism. Very few RFLPs have been sequenced to determine what sequence variation is responsible for the polymorphism. However, it may be problematic to interpret complex RFLP allelic systems in the absence of sequence information. RFLP analysis requires large amounts of high-quality DNA, has low genotyping throughput, and is very difficult to automate. Radioactive autography involving in genotyping and physical maintenance of RFLP probes limit its use and share between laboratories. RFLP markers were predominantly used in 1980s and 1990s, but since last decade fewer direct uses of RFLP markers in genetic research and plant breeding have been reported. Most plant breeders would think that RFLP is too laborious and demands too much pure DNA to be important for plant breeding. It was and is, however, central for various types of scientific studies.

RAPD Markers

RAPD is a PCR-based marker system. In this system, the total genomic DNA of an individual is amplified by PCR using a single, short (usually about ten nucleotides/bases) and random primer. The primer which binds to many different loci is used to amplify random sequences from a complex DNA template that is complementary to it (maybe including a limited number of mismatches). Amplification can take place during the PCR, if two hybridization sites are similar to one another (at least 3000 bp) and in opposite directions. The amplified fragments generated by PCR depend on the length and size of both the primer and the target genome. The PCR products (up to 3 kb) are separated by agarose gel electrophoresis and imaged by ethidium bromide (EB) staining. Polymorphisms resulted from mutations or rearrangements either at or between the primer-binding sites are visible in the electrophoresis as the presence or absence of a particular RAPD band.

RAPD predominantly provides dominant markers. This system yields high levels of polymorphism and is simple and easy to be conducted. First, neither DNA probes nor sequence information is required for the design of specific primers. Second, the procedure does not involve blotting or hybridization steps, and thus it is a quick, simple and efficient technique. Third, relatively small amounts of DNA (about 10 ng per reaction) are required and the procedure can be automated, and higher levels of polymorphism also can be detected compared with RFLP. Fourth, no marker development is required, and the primers are non-species specific and can be universal. Fifth, the RAPD products of interest can be cloned, sequenced and then converted into or used to develop other types of PCR-based markers, such as sequence characterized amplified region (SCAR), single nucleotide polymorphism (SNP), etc. However, RAPD also has some limitations/disadvantages, such as low reproducibility and incapability to detect allelic differences in heterozygotes.

AFLP Markers

AFLPs are PCR-based markers, simply RFLPs visualized by selective PCR amplification of DNA restriction fragments. Technically, AFLP is based on the selective PCR amplification of restriction fragments from a total double-digest of genomic DNA under high stringency conditions, i.e., the combination of polymorphism at restriction sites and hybridization of arbitrary primers. Because of this AFLP is also called selective restriction fragment amplification (SRFA). An AFLP primer (17-21 nucleotides in length) consists of a synthetic adaptor sequence, the restriction endonuclease recognition sequence and an arbitrary, non-degenerate 'selective' sequence (1-3 nucleotides). The primers used in this technique are capable of annealing perfectly to their target sequences (the adapter and restriction sites) as well as a small number of nucleotides adjacent to the restriction sites. The first step in AFLP involves restriction digestion of genomic DNA (about 500 ng) with two restriction enzymes, a rare cutter (6-bp recognition site, EcoRI, PtsI or HindIII) and a frequent cutter (4-bp recognition site, MseI or TaqI). The adaptors are then ligated to both ends of the fragments to provide known sequences for PCR amplification. The double-stranded oligonucleotide adaptors are designed in such a way that the initial restriction site is not restored after ligation. Therefore, only the fragments which have been cut by the frequent cutter and rare cutter will be amplified. This property of AFLP makes it very reliable, robust and immune to small variations in PCR amplification parameters (e.g., thermal cycles, template concentration), and it also can produce a high marker density. The AFLP products can be separated in high-resolution electrophoresis systems. The fragments in gel-based or capillary DNA sequencers can be detected by dye-labeling primers radioactively or fluorescently. The number of bands produced can be manipulated by the number of selective nucleotides and the nucleotide motifs used.

A typical AFLP fingerprint (restriction fragment patterns generated by the technique) contains 50-100 amplified fragments, of which up to 80% may serve as genetic markers. In general, AFLP assays can be conducted using relatively small DNA samples (1-100 ng per individual). AFLP has a very high multiplex ratio and genotyping throughput, and is relatively reproducible across laboratories. Another advantage is that it does not require sequence information or probe collection prior to generating the fingerprints, and a set of primers can be used for different species. This is especially useful when DNA markers are rare. However, AFLP assays have some limitations also. For instance, polymorphic information content for bi-allelic markers is low (the maximum is 0.5). High quality DNA is required for complete restriction enzyme digestion. AFLP markers usually cluster densely in centromeric regions in some species with large genomes (e.g., barley and sunflower). In addition, marker development is complicated and not cost-efficient, especially for locus-specific markers. The applications of AFLP markers include biodiversity studies, analysis of germplasm collections, and genotyping of individuals, identification of closely linked DNA markers, construction of genetic DNA marker maps, construction of physical maps, gene mapping, and transcript profiling.

SSR Markers

SSRs, also called microsatellites, short tandem repeat (STRs) or sequence-tagged microsatellite sites (STMS) are PCR-based markers. They are randomly tandem repeats of short nucleotide motifs (2-6 bp/nucleotides long). Di-, tri- and tetra-nucleotide repeats, e.g. (GT) n, (AAT) n and (GATA) n, are widely distributed throughout the genomes of plants and animals. The copy number of these repeats varies among individuals and is a source of polymorphism in plants. Because the DNA sequences flanking microsatellite regions are usually conserved, primers specific for these regions are designed for use in the PCR reaction. One of the most important attributes of microsatellite loci is their high level of allelic variation, thus making them valuable genetic markers. The unique sequences bordering the SSR motifs provide templates for specific primers to amplify the SSR alleles via PCR. SSR loci are individually amplified by PCR using pairs of oligonucleotide primers specific to unique DNA sequences flanking the SSR sequence. The PCR-amplified products can be separated in high-resolution electrophoresis systems (e.g. AGE and PAGE) and the bands can be visually recorded by fluorescent labeling or silver-staining.

SSR markers are characterized by their hyper-variability, reproducibility, co-dominant nature, locus-specificity, and random genome-wide distribution in most cases. The advantages of SSR markers include that they can be readily analyzed by PCR and easily detected by PAGE or AGE. SSR markers can be multiplexed, have high throughput genotyping and can be automated. SSR assays require only very small DNA samples (~100 ng per individual) and low start-up costs for manual assay methods. However, SSR technique requires nucleotide information for primer design, labor-intensive marker development process and high start-up costs for automated detections. Since the 1990s SSR markers have been extensively used in constructing genetic linkage maps, QTL mapping, marker-assisted selection and germplasm analysis in plants. In many species, plenty of breeder-friendly SSR markers have been developed and are available for breeders. For instance, there are over 35,000 SSR markers developed and mapped onto all 20 linkage groups in soybean.

SNP Markers

An SNP is a single nucleotide base difference between two DNA sequences or individuals. SNPs can be categorized according to nucleotide substitutions either as transitions (C/T or G/A) or transversions (C/G, A/T, C/A or T/G). In practice, single base variants in cDNA (mRNA) are considered to be SNPs as are single base insertions and deletions (indels) in the genome. SNPs provide the ultimate/simplest form of molecular markers as a single nucleotide base is the smallest unit of inheritance, and thus they can provide maximum markers. SNPs occur very commonly in animals and plants. Typically, SNP frequencies are in a range of one SNP every 100-300 bp in plants. SNPs may present within coding sequences of genes, non-coding regions of genes or in the intergenic regions between genes at different frequencies in different chromosome regions.

Based on various methods of allelic discrimination and detection platforms, many SNP genotyping methods have been developed. A convenient method for detecting SNPs is RFLP (SNP-RFLP) or by using the CAPS marker technique. If one allele contains a recognition site for a restriction enzyme while the other does not, digestion of the two alleles will produce different fragments in length. A simple procedure is to analyze the sequence data stored in the major databases and identify SNPs. Four alleles can be identified when the complete base sequence of a segment of DNA is considered and these are represented by A, T, G and C at each SNP locus in that segment. There are several SNP genotyping assays, such as allele-specific hybridization, primer extension, oligonucleotide ligation and invasive cleavage based on the molecular mechanisms, and different detection methods to analyze the products of each type of allelic discrimination reaction, such as gel electrophoresis, mass spectrophotometry, chromatography, fluorescence polarization, arrays or chips, etc. At the present, SNPs are also widely detected by sequencing.

SNPs are co-dominant markers, often linked to genes and present in the simplest/ultimate form for polymorphism, and thus they have become very attractive and potential genetic markers in genetic study and breeding. Moreover, SNPs can be very easily automated and quickly detected, with a high efficiency for detection of polymorphism. Therefore, it can be expected that SNPs will be increasingly used for various purposes, particularly as whole DNA sequences become available for more and more species (e.g., rice, soybean, maize, etc.). However, high costs for start-up or marker development, high-quality DNA required and high technical/equipment demands limit, to some extent, the application of SNPs in some laboratories and practical breeding programs.

The advantages or disadvantages of a marker system are relevant largely to the purposes of research, available genetic resources or databases, equipment and facilities, funding and personnel resources, etc. The choice and use of DNA markers in research and breeding is still a challenge for plant breeders. A number of factors need to be considered when a breeder chooses one or more molecular marker types . A breeder should make an appropriate choice that best meets the requirements according to the conditions and resources available for the breeding program.

Table: Comparison of most widely used DNA marker systems in plants.

Feature and description	RFLP	RAPD	AFLP	SSR	SNP
Genomic abundance	High	High	High	Moderate to high	Very high
Genomic coverage	Low copy coding region	Whole genome	Whole genome	Whole genome	Whole genome
Expression/inheritance	Co-dominant	Dominant	Dominant / co-dominant	Co-dominant	Co-dominant
Number of loci	Small (<1,000)	Small (<1,000)	Moderate (1,000s)	High (1,000s – 10,000s)	Very high (>100,000)
Level of polymorphism	Moderate	High	High	High	High

Type of polymorphism	Single base changes, indels	Single base changes, indels	Single base changes, indels	Changes in length of repeats	Single base changes, indels
Type of probes/primers	Low copy DNA or cDNA clones	10 bp random nucleotides	Specific sequence	Specific sequence	Allele-specific PCR primers
Cloning and/or sequencing	Yes	No	No	yes	yes
PCR-based	Yes	No	No	Yes	Yes
Radioactive detection	Usually yes	No	Yes or no	Usually no	No
Reproducibility/reliability	High	Low	High	High	High
Effective multiplex ratio	Low	Moderate	High	High	Moderate to high
Marker index	Low	Moderate	Moderate to high	High	Moderate
Genotyping throughput	Low	Low	High	High	High
Amount of DNA required	Large (5 – 50 μg)	Small (0.01 – 0.1 μg)	Moderate (0.5 – 1.0 μg)	Small (0.05 – 0.12 μg)	Small (≥ 0.05 μg)
Quality of DNA required	High	Moderate	High	Moderate to high	High
Technically demanding	Moderate	Low	Moderate	Low	High
Time demanding	High	Low	Moderate	Low	Low
Ease of use	Not easy	Easy	Moderate	Easy	Easy
Ease of automation	Low	Moderate	Moderate to high	High	High
Development/start-up cost	Moderate to high	Low	Moderate	Moderate to high	High
Cost per analysis	High	Low	Moderate	Low	Low
Number of polymorphic loci per analysis	1.0-3.0	1.5-5.0	20-100	1.0-3.0	1.0
Primary application	Genetics	Diversity	Diversity and genetics	All purposes	All purposes

Pre-requisites and General Activities of Marker-assisted Breeding

Prerequisites for an Efficient Marker-assisted Breeding Program

Compared with conventional breeding approaches, molecular breeding mainly referred to as DNA marker-assisted breeding, needs more complicated equipment and facilities. In general, the pre-requisites listed below are essential for marker-assisted breeding (MAB) in plants.

Appropriate marker system and reliable markers: For a plant species or crop, a suitable marker system and reliable markers available are critically important to initiate a marker-assisted breeding program. Suitable markers should have following attributes:

- Ease and low-cost of use and analysis.
- Small amount of DNA required.
- Co-dominance.

- Repeatability/reproducibility of results.
- High levels of polymorphism.
- Occurrence and even distribution genome wide.

In addition, another important desirable attribute for the markers to be used is close association with the target gene(s). If the markers are located in close proximity to the target gene or present within the gene, selection of the markers will ensure the success in selection of the gene. Although they can also be used in plant breeding programs, the number of classical markers possessing these features is very small. DNA markers for polymorphism are available throughout the genome, and their presence or absence is not affected by environments and usually do not directly affect the phenotype. DNA markers can be detected at any stage of plant growth, but the detection of classical markers is usually limited to certain growth stages. Therefore, DNA markers are the predominant types of genetic markers for MAB. Each type of markers has advantages and disadvantages for specific purposes. Relatively speaking, SSRs have most of the desirable features and thus are the current marker of choice for many crops. SNPs require more detailed knowledge of the specific, single nucleotide DNA changes responsible for genetic variation among individuals. However, more and more SNPs have become available in many species, and thus they are also considered an important type for marker-assisted breeding.

Quick DNA extraction and high throughput marker detection: For most plant breeding programs, hundreds to thousands of plants/individuals are usually screened for desired marker patterns. In addition, the breeders need the results instantly to make selections in a timely manner. Therefore, a quick DNA extraction technique and a high throughput marker detection system are essentially required to handle a large number of tissue samples and a large-scale screening of multiple markers in breeding programs. Extracting DNA from small tissue samples in 96- or 384-well plates and streamlined operations are adopted in many labs and programs. High throughput PAGE and AGE systems are commonly used for marker detection. Some labs also provide marker detection services using automated detection systems, e.g. SNP chips based on thousands to ten thousands of markers.

Genetic Maps

Linkage maps provide a framework for detecting marker-trait associations and for choosing markers to use in marker-assisted breeding. Therefore, a genetic linkage map, particularly high-density linkage map is very important for MAB. To use markers and select a desired trait present in a specific germplasm line, a proper population of segregation for the trait is required to construct a linkage map. Once a marker or a few markers are found to be associated with the trait in a given population, a dense molecular marker map in a standard reference population will help identify makers that are close to (or flank) the target gene. If a region is found associated with the desired traits of interest, fine mapping also can be done with additional markers to identify the marker(s) tightly linked to the gene controlling the trait. A favorable genetic map should have an adequate number of evenly-spaced polymorphic markers to accurately locate desired QTLs/genes.

Knowledge of marker-trait association: The most crucial factor for marker-assisted breeding is the knowledge of the associations between markers and the traits of interest. Only those markers that are closely associated with the target traits or tightly linked to the genes can provide sufficient guarantee for the success in practical breeding. The more closely the markers are associated with the traits, the higher the possibility of success and efficiency of use will be. This information can be obtained in various ways, such as gene mapping, QTL analysis, association mapping, classical mutant analysis, linkage or recombination analysis, bulked segregant analysis, etc. In addition, it is also critical to know the linkage situation, i.e. the markers are linked in cis/trans (coupling or repulsion) with the desired allele of the trait. Even if some markers have been reported to be tightly linked with a QTL, a plant breeder still needs to determine the association of alleles in his own breeding material. This makes QTL information difficult to directly transfer between different materials.

Quick and Efficient Data Processing and Management

In a marker-assisted breeding program, not only are large numbers of samples handled, but multiple markers for each sample also need to be screened at the same time. This situation requires an efficient and quick system for labeling, storing, retrieving, processing and analyzing large data sets, and even integrating data sets available from other programs. The development of bioinformatics and statistical software packages provides a useful tool for this purpose.

Activities of Marker-assisted Breeding

Marker-assisted breeding involves the following activities provided the prerequisites are well equipped or available:

- Planting the breeding populations with potential segregation for traits of interest or polymorphism for the markers used.
- Sampling plant tissues, usually at early stages of growth, e.g. emergence to young seedling stage.
- Extracting DNA from tissue sample of each individual or family in the populations, and preparing DNA samples for PCR and marker screening.
- Running PCR or other amplifying operation for the molecular markers associated with or linked to the trait of interest.
- Separating and scoring PCR/amplified products, by means of appropriate separation and detection techniques, e.g. PAGE, AGE, etc.
- Identifying individuals/families carrying the desired marker alleles.
- Selecting the best individuals/families with both desired marker alleles for target traits and desirable performance/phenotypes of other traits, by jointly using marker results and other selection criteria.

Depending upon the association between the markers and the traits as well as the status of marker alleles (homozygous or heterozygous), and advancing the individuals selected in breeding program until stable superior or elite lines that have improved traits are developed.

Marker-assisted Selection

MAS Procedure and Theoretical and Practical Considerations

Marker-assisted selection (MAS) refers to such a breeding procedure in which DNA marker detection and selection are integrated into a traditional breeding program. Taking a single cross as an example, the general procedure can be described as follow:

- Select parents and make the cross, at least one (or both) possesses the DNA marker allele(s) for the desired trait of interest.
- Plant F_1 population and detect the presence of the marker alleles to eliminate false hybrids.
- Plant segregating F_2 population, screen individuals for the marker(s), and harvest the individuals carrying the desired marker allele(s).
- Plant $F_{2:3}$ plant rows, and screen individual plants with the marker(s). A bulk of F_3 individuals within a plant row may be used for the marker screening for further confirmation in case needed if the preceding F_2 plant is homozygous for the markers. Select and harvest the individuals with required marker alleles and other desirable traits.
- In the subsequent generations (F_4 and F_5), conduct marker screening and make selection similarly as for $F_{2:3}$s, but more attention is given to superior individuals within homozygous lines/rows of markers.
- In $F_{5:6}$ or $F_{4:5}$ generations, bulk the best lines according to the phenotypic evaluation of target trait and the performance of other traits, in addition to marker data.
- Plant yield trials and comprehensively evaluate the selected lines for yield, quality, resistance and other characters of interest.

Theoretically, all the QTLs contributing to the trait of interest could be taken into account. For a quantitatively-inherited character like yield, numerous QTLs or genes are usually involved. It is almost impossible to select all QTLs or genes simultaneously so that the selected individuals incorporate all the desired QTLs due to the limitation of resources and facilities. The number of individuals in the population increases exponentially with the increase of target loci involved. The relative efficiency of MAS decreases as the number of QTLs increases and their heritability decreases. In other words, MAS will be less effective for a highly complex character governed by many genes than for a simply inherited character controlled by a few genes. The number of genes/QTLs not only impacts the efficiency of MAS, but also the breeding design and implement scheme. Typically no more than three QTLs are regarded as an appropriate and feasible choice, although five QTLs were used in improvement of fruit quality traits in tomato via marker-assisted introgression. With development of SNP markers (especially rapid automated detection and genotyping technologies), selection of more QTLs at the same time might be preferred and practicable.

For MAS for multiple genes/QTLs, it was suggested to limit the number of genes undergoing selection to three to four if they are QTLs selected on the basis of linked markers, and to five to six if they are known loci selected directly. Only the multi-environmentally verified QTLs that possess medium to large effects are selected. The first priority should be given to the major QTLs that can explain greatest proportion of phenotypic variation and/or can be consistently detected across a range of environments and different populations. In addition, an index for selection that weights markers differently could be constructed, depending on their relative importance to the breeding objectives.

The more markers associated with a QTL are used, the greater opportunity of success in selecting the QTL of interest will be ensured. However, efficiency is also important for a breeding program, especially when the resources and facilities are limited. From the point of both effectiveness and efficiency, for a single QTL it is usually suggested to use two markers (i.e. flanking markers) that are tightly linked to the QTL of interest. The markers to be used should be close enough to the gene/QTL of interest (<5cM) in order to ensure that only a minor proportion of the selected individuals will be recombinants. If a marker (e.g. the peak marker) is found to be located within the region of gene sequence of interest or in such a close proximity to the QTL/gene that no recombination occurs between the marker and the QTL/gene, such a marker only should be preferable. However, if a marker is not tightly linked to a gene of interest, recombination between the marker and gene may reduce the efficiency of MAS because a single crossover may alternate the linkage association and leads to selection errors. The efficiency of MAS decreases as the recombination frequency (genetic distance) between the marker and gene increases. Use of two flanking markers rather than one may decrease the chance of such errors due to homologous recombination and increase the efficiency of MAS. Only a double crossover (i.e. two single crossovers occurring simultaneously on both sides of the gene/QTL in the region) may result in selection errors, but the frequency of a double crossover is considerably rare. For instance, if two flanking markers with an interval of 20cM or so between them are used, there will be higher probability (99%) for recovery of the target gene than only one marker used.

In practical MAS, a breeder is also concerned about how the markers should be detected, how many generations of MAS have to be conducted, and how large size of the population is needed. In general, detection of marker polymorphism is performed at early stages of plant growth. This is true especially for marker-assisted backcrossing and marker-assisted recurrent selection, because only the individuals that carry preferred marker alleles are expected to be used in backcrossing to the recurrent parent and/or inter-mating between selected individuals/progenies. The generations of MAS required vary with the number of markers used, the degree of association between the markers and the QTLs/genes of interest, and the status of marker alleles. In many cases, marker screening is performed for two to four consecutive generations in a segregating population. If fewer markers are used and the markers are in close proximity to the QTL or gene of interest, fewer generations are needed. If homozygous status of marker alleles of interest is detected in two consecutive generations, marker screening may not be performed in their progenies. Bonnett et al., discussed the strategies for efficient implementation of MAS involving several issues, e.g. breeding systems or schemes, population sizes, number of target loci, etc. Their strategies include F_2 enrichment, backcrossing, and inbreeding.

In MAS, phenotypic evaluation and selection is still very helpful if conditions permit to do so, and even necessary in cases when the QTLs selected for MAS are not so stable across environments and the association between the selected markers and QTLs is not so close. Moreover, one should also take the impact of genetic background into consideration. The presence of a QTL or marker does not necessarily guarantee the expression of the desired trait. QTL data derived from multiple environments and different populations help a better understanding of the interactions of QTL x environment and QTL x QTL or QTL x genetic background, and thus help a better use of MAS. In addition to genotypic (markers) and phenotypic data for the trait of interest, a breeder often pays considerable attention to other important traits, unless the trait of interest is the only objective of breeding.

There are several indications for adoption of molecular markers in the selection for the traits of interest in practical breeding. The situations favorable for MAS include:

- The selected character is expressed late in plant development, like fruit and flower features or adult characters with a juvenile period (so that it is not necessary to wait for the plant to become fully developed before propagation occurs or can be arranged).
- The target gene is recessive (so that individuals which are heterozygous positive for the recessive allele can be selected and/or crossed to produce some homozygous offspring with the desired trait).
- Special conditions are required in order to invoke expression of the target gene(s), as in the case of breeding for disease and pest resistance (where inoculation with the disease or subjection to pests would otherwise be required), or the expression of target genes is highly variable with the environments.
- The phenotype of a trait is conditioned by two or more unlinked genes. For example, selection for multiple genes or gene pyramiding may be required to develop enhanced or durable resistance against diseases or insect pests.

MAS for Major Genes Or Improvement of Qualitative Traits

In crop plants, many economically important characteristics are controlled by major genes/QTLs. Such characteristics include resistance to diseases/pests, male sterility, self-incompatibility and others related to shape, color and architecture of whole plants and/or plant parts. These traits are often of mono- or oligogenic inheritance in nature. Even for some quality traits, one or a few major QTLs or genes can account for a very high proportion of the phenotypic variation of the trait. Transfer of such a gene to a specific line can lead to tremendous improvement of the trait in the cultivar under development. The marker loci which are tightly linked to major genes can be used for selection and are sometimes more efficient than direct selection for the target genes. In some cases, such advantages in efficiency may be due to higher expression of the marker mRNA in such cases that the marker is actually within a gene. Alternatively, in such cases that the target gene of interest differs between two alleles by a difficult-to-detect SNP, an external marker of which polymorphism is easier to detect, may present as the most realistic option.

Soybean cyst nematode (SCN) (Heterodera glycines Inchinoe) may be taken as an example of MAS for major genes. This pathogen is the most economically significant soybean pest. The principal strategy to reduce or eliminate damage from this pest is the use of resistant cultivars. However, identifying resistant segregants in breeding populations is a difficult and expensive process. A widely used phenotypic assay takes five weeks, requires a large greenhouse space, and about 5 to 10 h of labor for every 100 plant samples processed. Fortunately, the SSR marker Satt309 has been identified to be located only 1–2 cM away from the resistance gene rhg1, which forms the basis of many public and commercial breeding efforts. In a direct comparison, genotypic selection with Satt309 was 99% accurate in predicting lines that were susceptible in subsequent greenhouse assays for two test populations, and 80% accurate in a third population, each with a different source of SCN resistance. In soybean, Shi et al., reported that using molecular markers in a cross J05 x V94-5152, they developed five $F_{4:5}$ lines that were homozygous for all eight marker alleles linked to the genes/loci of resistance to soybean mosaic virus (SMV). These lines exhibited resistance to SMV strains G1 and G7 and presumably carried all three resistance genes (Rsv1, Rsv3 and Rsv4) that would potentially provide broad and durable resistance to SMV.

MAS for Improvement of Quantitative Traits

Most of the important agronomic traits are polygenic or controlled by multiple QTLs. MAS for the improvement of such traits is a complex and difficult task because it is related to many genes or QTLs involved, QTL x E interaction and epistasis. Usually, each of these genes has a small effect on the phenotypic expression of the trait and expression is affected by environmental conditions. Phenotyping of quantitative traits becomes a complex endeavor consequently, and determining marker-phenotype association becomes difficult as well. Therefore, repeated field tests are required to accurately characterize the effects of the QTLs and to evaluate the stability across environments. The QTL x E interaction reduces the efficiency of MAS and epistasis can result in a skewed QTL effect on the trait.

Despite a tremendous amount of QTL mapping experiments over the past decade, application and utilization of the QTL mapping information in plant breeding has been constrained by a number of factors:

- Strong QTL-environmental interaction which make phenotyping difficult since expression may vary from one location/year to another;
- Lack of universally valid QTL-marker associations applicable across populations. The notion that QTL mapping to identify new QTL markers whenever a new germplasm is used, puts some people off and they lose interest in MAS;
- Deficiencies in QTL statistical analysis which lead to either overestimation or underestimation of the number of QTLs involved and their effect on the trait;
- Often times, there are no QTLs with major effects on the trait and this means a large number of QTLs have to be identified and in many cases this becomes a tough goal to achieve and further complicates identification of marker-QTL association.

In order to improve the efficiency of MAS for quantitative traits, appropriate field experimental designs and approaches have to be employed. Attention should be given to replications both over time and space, consistency in experimental techniques, samplings and evaluations, robust data processing and statistical analysis. For example, composite interval mapping (CIM) allows the integration of data from different locations for joint analysis to estimate QTL-environment interaction so that stable QTLs across environments can be identified. A saturated linkage map enables accurate identification of both targeted QTLs as well as linked QTLs in coupling and repulsion linkage phases. In practical breeding for improvement of a quantitative trait, usually not many minor QTLs are considered but only a few major QTLs are used in MAS. In case many QTLs especially minor-effect QTLs are involved, a breeder would prefer to consider the strategy of gene pyramiding.

Fusarium head blight (FHB) caused by Fusarum species is one of the most destructive diseases in wheat and barley worldwide. To combat this disease, a great effort from multiple fields, including plant breeding and genetics, molecular genetics and genomics, plant pathology, and integrated management, has been dedicated since 1990s. Resistance to HFB in both wheat and barley is quantitatively inherited, and many QTLs have been identified from different resources of germplasm. Use of MAS to improve the resistance has become a choice for many breeding programs. In wheat, a major QTL designated as *Fhb1* was consistently detected across multiple environments and populations, and explained 20-40% of phenotypic variation in most cases. Thus wheat breeders would especially prefer to use this major QTL to develop new cultivars with FHB resistance. Pumphrey et al., compared 19 pairs of NIL for *Fhb1* derived from an ongoing breeding program and found that the average reduction in disease severity between NIL pairs was 23% for disease severity and 27% for kernel infection. Later investigation from the group also demonstrated successful implementation of MAS for this QTL.

Miedaner et al. demonstrated that MAS for three FHB resistance QTLs simultaneously was highly effective in enhancing FHB resistance in German spring wheat. FHB resistance was the highest in recombinant lines with multiple QTLs combined especially 3B plus 5A. Jiang et al. made a comparison of multiple-locus combinations in a RIL population derived from the cross "Veery x CJ 9306". For three loci, the average levels of resistance from low to high in genotypes were: no favorable allele – one favorable allele – two favorable alleles – three favorable alleles, except for the non-reciprocal comparisons. When four or five loci carrying favorable alleles from the resistant parent CJ 9306 were considered simultaneously, the coefficients of determination between the accumulated effects of alleles for different combinations and the averages of number or percentage of diseased spikelets for the corresponding RILs were 0.33-0.41 ($P<0.01$).

Marker-assisted Backcrossing

MABC Procedure and Theoretical and Practical Considerations

Marker-assisted or marker-based backcrossing (MABC) is regarded as the simplest form of marker-assisted selection, and at the present it is the most widely and successfully used method in practical molecular breeding. MABC aims to transfer one or a few genes/QTLs of interest from one genetic source (serving as the donor parent and maybe inferior agronomically or not good enough in comprehensive performance in many cases) into a superior cultivar or elite breeding line (serving as the recurrent parent) to improve the targeted trait. Unlike traditional backcrossing, MABC is based on the alleles of markers associated with or linked to gene(s)/QTL(s) of interest instead of phenotypic performance of target trait. The general procedure of MABC is as follow, regardless of dominant or recessive nature of the target trait in inheritance:

- Select parents and make the cross, one parent is superior in comprehensive performance and serves as recurrent parent (RP), and the other one used as donor parent (DP) should possess the desired trait and the DNA markers allele(s) associated with or linked to the gene for the trait.

- Plant F_1 population and detect the presence of the marker allele(s) at early stages of growth to eliminate false hybrids, and cross the true F_1 plants back to the RP.
- Plant BCF_1 population, screen individuals for the marker(s) at early growth stages, and cross the individuals carrying the desired marker allele(s) (in heterozygous status) back to the RP. Repeat this step in subsequent seasons for two to four generations, depending upon the practical requirements and operation situations.
- Plant the final backcrossing population (e.g. BC_4F_1), and screen individual plants with the marker(s) for the target trait and discard the individuals carrying homozygous markers alleles from the RP. Have the individuals with required marker allele(s) selfed and harvest them.
- Plant the progenies of backcrossing-selfing (e.g. BC_4F_2), detect the markers and harvest individuals carrying homozygous DP marker allele(s) of target trait for further evaluation and release.

Theoretically, the proportion of the RP genome after n generations of backcrossing is given by $1 - (1/2)^{n+1}$ for a single locus and $[1 - (1/2)^{n+1}]^k$ for k loci, respectively, for a population large enough in size (or with adequate individuals) and no selection being made during backcrossing (i.e. "blind" backcrossing only). The percentage of the RP genome is the average of the population, with some individuals possessing more of the RP genome than others. To fully recover the genome of the RP, 6-8 generations of backcrossing is needed typically in case no selection is made for the RP. However, this process is usually slower than expected for the target gene-carrier chromosome, i.e. linkage drag, especially in case a linkage exists between the target gene and other undesirable traits. On the other hand, the process of introgression of QTLs/genes and recovery of the RP genome may be accelerated by selection using markers flanking QTLs and evenly spaced markers from other chromosomes (i.e. unlinked to QTLs) of the RP or selection for the performance of the RP conducted simultaneously. For MABC program, therefore, there are two types of selection recognized: Foreground selection and background selection.

In foreground selection, the selection is made only for the marker allele(s) of donor parent at the target locus to maintain the target locus in heterozygous state until the final backcrossing is completed. Then the selected plants are selfed and the progeny plants with homozygous DP allele(s) of selected markers are harvested for further evaluation and release. The effectiveness of foreground selection depends on the number of genes/loci involved in the selection, the marker-gene/QTL association or linkage distance and the undesirable linkage to the target gene/QTL.

In background selection, the selection is made for the marker alleles of recurrent parent in all genomic regions of desirable traits except the target locus, or selection against the undesirable genome of donor parent. The objective is to hasten the restoration of the RP genome and eliminate undesirable genes introduced from the DP. The progress in recovery of the RP genome depends on the number of markers used in background selection. The more markers evenly located on all the chromosomes are selected for the RP alleles, the faster recovery of the RP genome will be achieved but larger population size and more genotyping will be required as well. In addition, the linkage drag also can be efficiently addressed by background selection using DNA markers, although it is difficult to overcome in a traditional backcrossing program.

Foreground selection and background selection are two respective aspects of MABC with different foci of selection. In practice, however, both foreground and background selection is usually conducted in the same program, either simultaneously or successively. In many cases, they can be performed alternatively even in the same generation. The individuals that have the desired marker alleles for target trait are selected first (foreground selection). Then the selected individuals are screened for other marker alleles again for the RP genome (background selection). It is understandable to do so because selection of the target gene/QTL is the essential and only critical point for backcrossing program, and the individuals that do not have the allele of target gene will be discarded and thus it is not necessary to genotype them for other traits.

The efficiency of MABC depends upon several factors, such as the population size for each generation of backcrossing, marker-gene association or the distance of markers from the target locus, number of markers used for target trait and RP background, and undesirable linkage drag. Based on simulations of 1000 replicates, Hospital presented the expected results of a typical MABC program, in which heterozygotes were selected at the target locus in each generation, and RP alleles were selected for two flanking markers on target chromosome each located 2 cM apart from the target locus and for three markers on non-target chromosomes. As shown in Table below, a faster recovery of the RP genome could be achieved by MABC with combined foreground and background selection, compared to traditional backcrossing. Therefore, using markers can lead to considerable time savings compared to conventional backcrossing.

Table: Expected results of a MABC program with combined foreground and background selection used.

Backcross generation	Number of individuals	% homozygosity of recurrent parent alleles at selected markers		% recurrent parent genome	
		Chromosome with target locus	All other chromosomes	Marker-assisted backcross	Conventional backcross
BC_1	70	38.4	60.6	79.0	75.0
BC_2	100	73.6	87.4	92.2	87.5
BC_3	150	93.0	98.8	98.0	93.7
BC_4	300	100.0	100.0	99.0	96.9

In a MABC program, the population to be analyzed should contain at least one genotype that has all favorable alleles for a particular QTL. Later, the number of QTLs may be increased progressively, but not beyond six QTLs in most cases because of prohibitive difficulty in handling all QTLs. In addition, the more QTLs/genes are transferred, the larger the proportion of unwanted genes would be due to linkage drag. In general, most of the unwanted genes are located on non-target chromosomes in early BC generations, and are rapidly removed in subsequent BC generations. On the contrary, the quantity of DP genes on the target chromosome decreases much more slowly, and even after generation BC6 many of the unwanted donor genes are still located on the target chromosome in segregating state. Given a total genome length is 3000 cM, 1% donor DNA fragments after six backcrosses represents a 30 cM chromosomal segment or region, which may host many unwanted genes, especially if the DP is a wild genetic resource. Young and Tanksley genotyped a collection of tomato varieties in which the resistance gene was previously transferred at the Tm-2 locus with RLFP markers. Their data indicated that the size of chromosomal segment retained around the Tm-2 locus during backcross breeding was very variable, with one line exhibiting a donor segment of 50 cM after 11 backcrosses and other one possessing 36 cM donor segment after 21 backcrosses. This clearly demonstrates the need for background selection.

Linkage drag can be reduced by performing background selection. Typically, two markers flanking the target gene are used, and the individuals (or double recombinants) that are heterozygous at the target locus and homozygous for the recipient (RP) alleles at both flanking markers are selected. Use of closer flanking markers leads to more effective and faster reduction of linkage drag compared to distant markers. However, less distance between two flanking markers implies less probability of double recombination, and thus larger populations and more genotyping are needed. In order to optimize genotyping effort (i.e. the cost of the program), therefore, it is important to determine the minimal population sizes necessary to ensure the desired genotypes can be obtained. Hospital and Decoux developed statistical software for determining the minimum population size required in BC program to identify at least one individual that is double-recombinant with heterozygosity at target locus and homozygosity for recurrent parent alleles at flanking marker loci. In addition, for closely-linked flanking markers, it is unlikely to obtain double recombinant genotypes through only one generation of backcrossing. Therefore, additional backcrossing should be conducted. For instance, in one BC generation (e.g. BC1) single recombination on one side of the target gene is selection, and single recombination on the other side may be selected in another BC generation (e.g. BC2). In this way, individuals with desired RP alleles at two flanking markers and donor allele at target locus can be finally obtained.

To accelerate the recovery of RP genome on non-target chromosomes, scientists suggested using markers in backcrossing and discussed how many makers should be used. In background selection, the approaches involve selecting individuals that are of homozygous recipient type at a collection of markers located on non-carrier chromosomes. From a point of both effectiveness and efficiency, it is important to determine an appropriate number of markers to be used. More markers do not necessarily mean better benefits in practice. Generally, several markers are involved and MABC should be performed over two or more generations. It is unlikely that the selection objective can be realized in a single BC generation.

Dense marker coverage of non-target chromosomes is not mandatory to increase the overall proportion of recurrent parent genome, unless fine-mapping of specific chromosome regions is highly important. An appropriate number of markers and optimal position on chromosomes are important. Computer simulation suggested that for a chromosome of 100 cM, two to four markers are sufficient, and selection based on markers would be most efficient if the markers are optimally positioned along the chromosomes. In practice, at least two or three markers per chromosome are needed, and every chromosome should be involved. In such a MABC scheme, three to four generations of backcrossing is generally enough to achieve more than 99% of the recurrent parent genome. With respect to the time necessary to release new varieties, the gain due to background selection can be economically valuable. In addition, background selection is more efficient in late BC generations than in early BC generations. For example, if a BC breeding scheme is conducted over three successive BC generations and yet the preference is to genotype individuals only once, then it is more efficient to genotype and select the individuals in BC3 generation rather than in the BC1 generation.

Application of MABC

Success in integrating MABC as a breeding approach lies in identifying situations in which markers offer noticeable advantages over conventional backcrossing or valuable complements to conventional breeding effort. MABC is essential and advantageous when:

- Phenotyping is difficult and/or expensive or impossible.
- Heritability of the target trait is low.
- The trait is expressed in late stages of plant development and growth, such as flowers, fruits, seeds, etc.
- The traits are controlled by genes that require special conditions to express.
- The traits are controlled by recessive genes.
- Gene pyramiding is needed for one or more traits.

Among the molecular breeding methods, MABC has been most widely and successfully used in plant breeding up to date. It has been applied to different types of traits (e.g. disease/pest resistance, drought tolerance and quality) in many species, e.g. rice, wheat, maize, barley, pear millet, soybean, tomato, etc. In maize, for example, Bacillus thuringiens is a bacterium that produces insecticidal toxins, which can kill corn borer larvae when they ingest the toxins in corn cells. The integration of the Bt transgene into various corn genetic backgrounds has been achieved by using MABC. Aroma in rice is controlled by a recessive gene which is due to an eight base-pair deletion and three single nucleotide polymorphism in a gene that codes for betaine aldehyde dehydrogenase 2. This discovery allows identification of the aromatic and non-aromatic rice varieties and discriminates homozygous recessive and dominant as well as heterozygous individuals in segregating population for the trait. MABC has been used to select for aroma in rice. High lysine opaque2 gene in corn was incorporated using MABC. However, the rate of success decreases when large numbers of QTLs are targeted for introgression. Sebolt et al. used MABC for two QTL for seed protein content in soybeans. However, only one QTL was confirmed in BC3F4:5. When that QTL was introduced in three different genetic backgrounds, it had no effect in one background. In tomato, Tanksley and Nelson proposed a MABC strategy, called advanced backcross-QTL (AB-QTL), to transfer resistance genes from wild relative/unadapted genotype into elite germplasm. The strategy has proven effective for various agronomically important traits in tomato, including fruit quality and black mold resistance. In addition, AB-QTL has been used in other crop species, such as rice, barley, wheat, maize, cotton and soybean, collectively demonstrating that this strategy is effective in transferring favorable alleles from the wild/unadapted germplasm to elite germplasm.

In barley, a marker linked (0.7 cM) to the Yd2 gene for resistance to barley yellow dwarf virus was successfully used to select for resistance in a backcrossing scheme. Compared to lines without the marker, the BC2F2-derived lines carrying the linked marker had lighter leaf symptoms and higher yield when infected by the virus. In maize, marker-facilitated backcrossing was also successfully employed to improve complex traits such as grain yield. Using MABC, six chromosomal segments each in two elite lines, Tx303 and Oh43, were transferred into two widely used inbred lines, B73 and Mo17, through three generations of backcrossing followed by two selfing generations. Then the enhanced lines with better performance were selected based on initial evaluations of testcross hybrids. The single-cross hybrids of enhanced B73 x enhanced Mo17 out-yielded the check hybrids by 12-15%. Zhao et al. reported that a major quantitative trait locus (named qHSR1) for resistance to head smut in maize was successfully integrated into ten high-yielding inbred lines (susceptible to head smut). Each of the ten high-yielding lines was crossed with a donor parent Ji 1037 that contains qHSR1 and is completely resistant to head smut, followed by five generations of back-crossing to the respective recurrent parents. In BC1 through BC3 only phenotypic selection was conducted to identify highly resistant individuals after artificial inoculation. In BC4phenotypic selection, foreground selection and recombinant selection were conducted to screen for resistant individuals with the shortest qHSR1 donor regions. In BC5, phenotypic selection, foreground selection and background selection were performed to identify resistant individuals with the highest proportion of the recurrent parent genome, followed by one generation of self-pollination to obtain homozygous genotypes at the qHSR1 locus. The ten improved inbred lines all showed substantial resistance to head smut, and the hybrids derived from these lines also showed a significant increase in the resistance.

Currently, a cooperative marker-based backcrossing project for high-oleic acid in soybean has been initiated among multiple U.S. land-grant universities and USDA-ARS. Backcrossing and selection will be performed using the markers tightly linked to the high-oleic genes/loci. Hopefully, the high-oleic (80% or higher) traits will be successfully transferred from mutant lines or derived lines into other locally superior cultivars/lines, or combined with other unique traits like low linolenic acid.

Marker-assisted Gene Pyramiding and Marker-assisted Recurrent Selection

Marker-assisted gene pyramiding (MAGP) is one of the most important applications of DNA markers to plant breeding. Gene pyramiding has been proposed and applied to enhance resistance to disease and insects by selecting for two or more than two genes at a time. For example in rice such pyramids have been developed against bacterial blight and blast reported a success in pyramiding qualitative gene and QTLs for resistance to stripe rust in barley. The advantage of using markers in this case allows selecting for QTL-allele-linked markers that have the same phenotypic effect. To enhance or improve a quantitatively inherited trait in plant breeding, pyramiding of multiple genes or QTLs is recommended as a potential strategy. The cumulative effects of multiple-QTL pyramiding have been proven in crop species like wheat, barley and soybean. Pyramiding of multiple genes/QTLs may be achieved through different approaches: multiple-parent crossing or complex crossing, backcrossing, and recurrent selection. A suitable breeding scheme for MAGP depends on the number of genes/QTLs required for improvement of traits, the number of parents that contain the required genes/QTLs, the heritability of traits of interest, and other factors (e.g. marker-gene association, expected duration to complete the plan and relative cost). Assuming three or four desired genes/QTLs exist separately in three or four lines, pyramiding of them can be realized by three-way, four-way or double crossing. They may also be integrated by convergent backcrossing or stepwise backcrossing. However, if there are more than four genes/QTLs to be pyramided complex or multiple crossing and/or recurrent selection may be often preferred.

For MABC-based gene pyramiding, in general, there may be three strategies or breeding schemes: stepwise, simultaneous/synchronized and convergent backcrossing or transfer. Supposing one cultivar W is superior in comprehensive performance but lack of a trait of interest, and four different genes/QTLs contributing to the trait have been identified in four germplasm lines (e.g. P1, P2, P3 and P4). Three MABC schemes for pyramiding the genes/QTLs can be described as follow:

- Scheme 1
  - Stepwise Backcrossing
- Scheme 2
  - Simultaneous/Synchronized Backcrossing
- Scheme 3
  - Convergent Backcrossing

In the stepwise backcrossing, four target genes/QTLs are transferred into the recurrent parent W in order. In one step of backcrossing, one gene/QTL is targeted and selected, followed by next step of backcrossing for another gene/QTL, until all target genes/QTLs have been introgressed into the RP. The advantage is that gene pyramiding is more precise and easier to implement as it involves only one gene/QTL at one time and thus the population size and genotyping amount will be small. The improved recurrent parent may be released before the final step as long as the integrated genes/QTLs (e.g. two or three) meet the requirement at that time. The disadvantage is that it takes a longer time to complete. In the simultaneous or synchronized backcrossing, the recurrent parent W is first crossed to each of four donor parents to produce four single-cross F_1s. Two of the four single-cross F_1s are crossed with each other to produce two double-cross F_1s, and these two double-cross F_1s are crossed again to produce a hybrid integrating all four target genes/QTLs in heterozygous state. The hybrid and/or progeny with heterozygous markers for all four target genes/QTLs is subsequently crossed back to the RP W until a satisfactory recovery of the RP genome, and finalized by one generation of selfing. The advantage of this method is that it takes the shortest time to complete. However, in the backcrossing all target genes/QTLs are involved at the same time and thus it requires a large population and more genotyping. Convergent backcrossing is a strategy combining the advantages of stepwise and synchronized backcrossing. First the four target gene/QTLs are transferred separately from the donors into the recurrent parent W by single crossing followed by backcrossing based on markers linked to the target genes/QTLs, to produce four improved lines (W^{AA}, W^{BB}, W^{CC}, and W^{DD}). Two of the improved lines are crossed with each other and the two hybrids are then intercrossed to integrate all four genes/QTLs together and develop the final improved line with all four genes/QTLs pyramided (i.g. $W^{AABBCCDD}$). Relatively speaking, convergent backcrossing is more acceptable because in this scheme not only is time reduced (compared to stepwise transfer) but gene fixation and/or pyramiding is also more easily assured (compared to simultaneous transfer).

Theoretical issues and efficiency of MABC for gene pyramiding have been investigated through computer simulations. Practical application of MABC to gene pyramiding has been reported in many crops, including rice, wheat, barley, cotton, soybean, common bean and pea, especially for developing durable resistance to stresses in crops. However, there is very limited information available about the release of commercial cultivars resulted from this strategy. Somers et al implemented a molecular breeding strategy to introduce multiple pest resistance genes into Canadian wheat. They used high throughput SSR genotyping and half-seed analysis to process backcrossing and selection for six FHB resistance QTLs, plus orange blossom wheat midge resistance gene Sm1 and leaf rust resistance gene Lr21. They also used 45-76 SSR markers to perform background selection in backcrossing populations to accelerate the restoration of the RP genetic background. This strategy resulted in 87% fixation of the elite genetic background at the BC_2F_1 on average and successfully introduced all (up to 4) of the chromosome segments containing FHB, Sm1 and *Lr21* resistance genes in four separate crosses.

Similar to the simultaneous/synchronized backcrossing scheme, marker-assisted complex or convergent crossing (MACC) can be undertaken to pyramid multiple genes/QTLs. In particular, MACC is a proper option of breeding schemes for gene pyramiding if all the parents are improved cultivars or lines with good comprehensive performance and have different or complementary genes or favorable alleles for the traits of interest. The difference from simultaneous backcrossing is that selfing hybrid and progenies replaces backcrossing hybrid to the recurrent parent. In MACC, the hybrid of convergent crossing is subsequently self-pollinated and marker-based selection for target traits is performed for several consecutive generations until genetically stable lines with desired marker alleles and traits have been developed. In order to reduce population size and to avoid loss of most important genes/QTLs, different markers may be used and selected in different generations, depending on their relative importance. The markers for the most important genes/QTLs can be detected and selected first in early generations and less important markers later. Once homozygous alleles of the markers for a gene/locus are detected, they may not be necessarily detected again in the subsequent generations. Instead, phenotypic evaluation should be conducted if conditions permit.

Using markers to select or pyramid for multiple genes/QTLs is more complex and less proven. Recurrent selection is widely regarded as an effective strategy for the improvement of polygenic traits. However, the effectiveness and efficiency of selection are not so satisfactory in some cases because phenotypic selection is highly dependent upon environments and genotypic selection takes a longer time (2-3 crop seasons at least for one cycle of selection). Marker-assisted recurrent selection (MARS) is a scheme which allows performing genotypic selection and intercrossing in the same crop season for one cycle of selection. Therefore, MARS could enhance the efficiency of recurrent selection and accelerate the progress of the procedure, particularly helps in integrating multiple favorable genes/QTLs from different sources through recurrent selection based on a multiple-parental population.

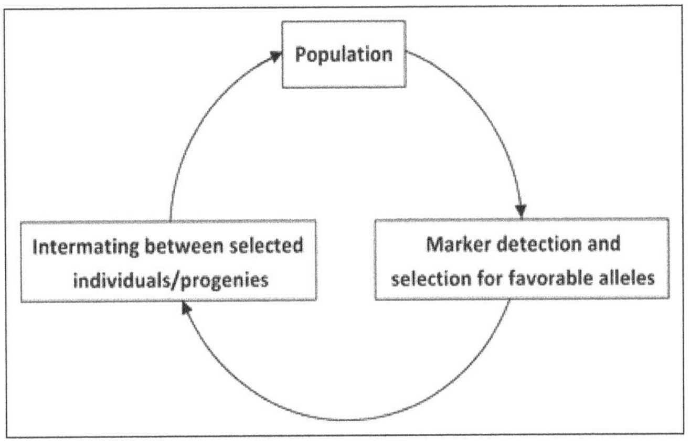

For complex traits such as grain yield, biotic and abiotic resistance, MARS has been proposed for "forward breeding" of native genes and pyramiding multiple QTLs. MARS is a recurrent selection scheme using molecular markers for the identification and selection of multiple genomic regions involved in the expression of complex traits to assemble the best-performing genotype within a single or across related populations. Johnson presented an example to demonstrate the efficiency of MARS for quantitative traits. In their maize MARS programs, a large-scale use of markers in bi-parental populations, first for QTL detection and then for MARS on yield (i.e. rapid cycles of recombination and selection based on associated markers for yield), could allow increased efficiency of long-term selection by increasing the frequency of favorable alleles. Eathington and Crosbie et al. also indicated that the genetic gain achieved through MARS in maize was about twice that of phenotypic selection (PS) in some reference populations. In upland cotton, Yi et al. reported significant effectiveness of MARS for resistance to Helicoverpa armigera. The mean levels of resistance in improved populations after recurrent selection were significantly higher than those of preceding populations.

Genomic Selection

Genomic selection (GS) or genome-wide selection (GWS) is a form of marker-based selection, which was defined by Meuwissen as the simultaneous selection for many (tens or hundreds of thousands of) markers, which cover the entire genome in a dense manner so that all genes are expected to be in linkage disequilibrium with at least some of the markers. In GS genotypic data (genetic markers) across the whole genome are used to predict complex traits with accuracy sufficient to allow selection on that prediction alone. Selection of desirable individuals is based on genomic estimated breeding value (GEBV), which is a predicted breeding value calculated using an innovative method based on genome-wide dense DNA markers. GS does not need significant testing and identifying a subset of markers associated with the trait. In other words, QTL mapping with populations derived from specific crosses can be avoided in GS. However, it does first need to develop GS models, i.e. the formulae for GEBV prediction. In this process (training phase), phenotypes and genome-wide genotypes are investigated in the training population (a subset of a population) to predict significant relationships between phenotypes and genotypes using statistical approaches. Subsequently, GEBVs are used for the selection of desirable individuals in the breeding phase, instead of the genotypes of markers used in traditional MAS. For accuracy of GEBV and GS, genome-wide genotype data is necessary and require high marker density in which all quantitative trait loci (QTLs) are in linkage disequilibrium with at least one marker.

GS can be possible only when high-throughput marker technologies, high-performance computing and appropriate new statistical methods become available. This approach has become feasible due to the discovery and development of large number of single nucleotide polymorphisms (SNPs) by genome sequencing and new methods to efficiently genotype large number of SNP markers. As suggested by Goddard and Hayes, the ideal method to estimate the breeding value from genomic data is to calculate the conditional mean of the breeding value given the genotype at each QTL. This conditional mean can only be calculated by using a prior distribution of QTL effects, and thus this should be part of the research to implement GS. In practice, this method of estimating breeding values is approximated by using the marker genotypes instead of the QTL genotypes, but the ideal method is likely to be approached more closely as more sequence and SNP data are obtained.

Since the application of GS was proposed by Meuwissen et al to breeding populations, theoretical, simulation and empirical studies have been conducted, mostly in animals. Relatively speaking, GS in plants was less studied and large-scale empirical studies are not available in public sectors for plant breeding, but it has attracted more and more attention in recent years. Studies indicated that in all cases, accuracies provided by GS were greater than might be achieved on the basis of pedigree information alone. In oil palm, for a realistic yet relatively small population, GS was superior to MARS and PS in terms of gain per unit cost and time. The studies have demonstrated the advantages of GS, suggesting that GS would be a potential method for plant breeding and it could be performed with realistic sizes of populations and markers when the populations used are carefully chosen.

GS has been highlighted as a new approach for MAS in recent years and is regarded as a powerful, attractive and valuable tool for plant breeding. However, GS has not become a popular methodology in plant breeding, and there might be a far way to go before the extensive use of GS in plant breeding programs. The major reason might be the unavailability of sufficient knowledge of GS for practical use. Statistics and simulation discussed in terms of formulae in GS studies are most likely too specific and hard for plant breeders to understand and to use in practical breeding programs. From a plant breeder's point of view, GS can be practicable for a few breeding populations with a specific purpose, but may be impractical for a whole breeding program dealing with hundreds and thousands of crosses/populations at the same time. Therefore, GS must shift from theory to practice, and its accuracy and cost effectiveness must be evaluated in practical breeding programs to provide convincing empirical evidence and warrant a practicable addition of GS to a plant breeder's toolbox. Development of easily understandable formulae for GEBVs and user-friendly software packages for GS analysis is helpful in facilitating and enhancing the application of GS in plant breeding.

Plant Mutagenesis

It is well known that evolution and practical breeding depend on genetic variation. The variations that are found in nature do not represent the original spectra of spontaneous mutations. Rather, they are the result of genotypes recombining within populations and their continuous interaction with environmental factors. Green plants are essential for human existence as a source of food, clothing and energy resources. As highlighted, prehistoric hunter-gatherers depended on their hunting skills and on abundant natural vegetation to get non-poisonous and nutritious fruits, tubers, seeds and other food stuffs. With the growth of human population, larger and safer food supplies had to be found and gradually large-scale production systems based on plant domestication were developed. The means by which new plant varieties were developed for cultivation and used by humans is called plant breeding. Primarily, simple selection of desirable offspring was the first method of breeding and this utilized the occurrence of spontaneous mutations.

Genetics became a fundamental science of plant breeding after Gregor Johann Mendel discovered the laws of heredity in the nineteenth century. However, further advancements in plant breeding took place when the hybridization methodology was developed. Its aim was to combine desirable genes found in two or more different varieties in order to produce pure-line progeny superior to the parental types in many respects. Cross breeding (or recombinant breeding), based on crossing of different genotypes followed by trait selection, has become a common practice in plant breeding. Later, the work on the induction of genetic alterations through X-rays by Lewis John Stadler in the late 1920s and early 1930s laid the foundation of another type of plant breeding known as mutation breeding. The variation so created is further amplified by recombination of alleles on homologous chromosomes and their independent assortment at meiosis. Mutations are the primary source of all genetic variations existing in any organism, including plants. The resulting variation provides the raw material for natural selection and is also a driving force in evolution. Spontaneous mutations are very rare and random in terms of time of occurrence, which makes them more difficult to use in plant breeding programmes. In this way, mutant forms showing both large and small effects on the phenotype arise for all kinds of traits. Mutation breeding involves the development of new varieties by generating and utilizing genetic variability through chemical and physical mutagenesis. It is now a pillar of modern plant breeding, along with recombinant breeding and transgenic breeding. As underlined by Novak and Brunner, this method, often supplemented by germplasm derived from induced mutation, has become the most common one for breeding plants through sexual reproduction.

In crops that do not produce seeds, e.g. edible banana or seedless grapes, mutation induction may be the only productive way of increasing variability for developing new cultivars. This also applies to many root and tuber crops, and the development of novel colours and variations in ornamental plant species propagated vegetatively. Crossbreeding in vegetatively propagated perennial crop species, such as many fruit crops, is also subject to constraints of time, growing space and clonal identity. In this case, mutation induction can be a valuable breeding strategy. For plants that are fully sterile without seeds, alternative approaches have to be developed. This is the technique of somatic tissue manipulation by mutation breeding and biotechnology.

Mutation Breeding

Mutagenesis is the process whereby sudden heritable changes occur in the genetic information of an organism not caused by genetic segregation or genetic recombination, but induced by chemical, physical or biological agents. Mutation breeding employs three types of mutagenesis. These are induced mutagenesis, in which mutations occur as a result of irradiation (gamma rays, X-rays, ion beam, etc.) or treatment with chemical mutagens; site-directed mutagenesis, which is the process of creating a mutation at a defined site in a DNA molecule; and insertion mutagenesis, which is due to DNA insertions, either through genetic transformation and insertion of T-DNA or activation of transposable elements. Plant breeding requires genetic variation of useful traits for crop improvements. However, multiple mutant alleles are the sources of genetic diversity for crop breeding as well as functional analysis of the targeted gene in many cases. The key point in mutation breeding is the process of identifying individuals with a target mutation, which involves two major steps: mutant screening and mutant confirmation. Mutant screening is a process involving selection of individuals from a large mutated population that meet specific selection criteria, e.g. early flowering, disease resistance as compared to the parent. However, these selections are often regarded as putative mutants or false mutants. Mutant confirmation, on the other hand, is the process of reevaluating the putative mutants under a controlled and replicated environment using large samples. Through this process, many putative mutants are revealed to be false mutants. In general, the mutations that are important in crop improvement usually involve single bases and may or may not affect protein synthesis.

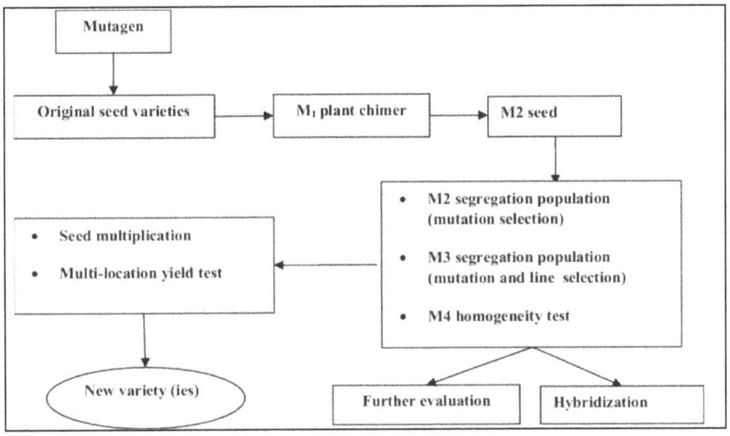

Mutagenic Agents

Agents of artificial mutations are called mutagens. They are generally grouped into two broad categories, namely chemical mutagens and physical mutagens. Traditionally, to induce mutations in crops, planting materials are exposed to physical and chemical mutagenic agents. Mutagenesis can be performed with all types of planting materials, e.g. whole plants, usually seedlings, and in vitro cultured cells. Nevertheless, the most commonly used plant material is seed. Multiple forms of plant propagules, such as bulbs, tubers, corms and rhizomes and more recently, the induction of mutations in vegetatively propagated plants is becoming more efficient as scientists take advantage of totipotency (ability of a single cell to divide and produce all of the differentiated cells in an organism to regenerate into whole plants) using single cells and other forms of in vitro cultured plant tissues. The starting materials for the induction of mutations are vegetative cuttings, scions, or in vitro cultured tissues like leaf and stem explants, anthers, calli, cell cultures, microspores, ovules, protoplasts, etc. Gametes, usually inside the inflorescences, are also targeted for mutagenic treatments through immersion of spikes, tassels, etc. Whereas chemical mutagens are preferably used to induce point mutations, physical mutagens induce gross lesions, such as chromosomal abbreviation or rearrangements. As pointed out by Mba, it is noteworthy that the frequency and types of mutations are direct results of the dosage and rate of exposure or administration of the mutagen rather than its type. In the end, the choice of a mutagen will be based more often than not on the particular researcher's circumstances, such as safety of usage, ease of use, availability of the mutagens, effectiveness in inducing certain genetic alterations, suitable tissue, cost and available infrastructure among other factors.

Physical Mutagenesis

In the past 80 years, physical mutagens, mostly ionizing radiations, have been used widely for inducing hereditary aberrations and more than 70% of mutant varieties were developed using physical mutagenesis. Radiation is defined as energy travelling through a distance in the form of waves or particles. These are relatively high-energy levels of electromagnetic (EM) spectrum that are capable of dislodging electrons from the nuclear orbits of the atoms that they impact upon. The impacted atoms, therefore, become ions. Hence, the term ionizing radiation. These ionizing components of the EM include cosmic, gamma (g) and X-rays. The most commonly used physical mutagens are shown in Table below. X-rays were the first to be used to induce mutations. Since then, various subatomic particles (neutrons, protons, beta particles and alpha particles) have been generated using nuclear reactors. Gamma radiation from radioactive cobalt (^{60}Co) is widely used. It has high penetrating potential and is hazardous.

Table: Examples of commonly used physical mutagens.

Mutagen	Source	Characteristics	Hazard
X-rays	X-ray machine	Electromagnetic radiation; penetrates tissues from a few millimeters to many centimeter	Dangerous, penetrating
Gamma ray	Radioisotopes and nuclear reaction	Electromagnetic radiation produced by radioisotopes and nuclear reactors; very penetrating into tissues; sources are ^{60}Co (Cobalt-60) and ^{137}Cs (Caesium-137)	Dangerous, very penetrating
Neutrons	Nuclear reactors or accelerators	There are different types (fast, slow, thermal); produced in nuclear reactors; uncharged particles; penetrate tissues to many centimeters; source is ^{235}U	Very hazardous
Beta particles	Radioactive isotopes or accelerators	Produced in particle accelerators or from radioisotopes; are electrons; ionize; shallowly penetrating; sources include ^{32}P and ^{14}C	May be dangerous
Alpha particles	Radioisotopes	Derived from radioisotopes; a helium nucleus capable of heavyionization; very shallowly penetrating	Very dangerous
Protons	Nuclear reactors or accelerators	Produced in nuclear reactors and accelerators; derived from hydrogen nucleus; penetrate tissues up to several centimeters	Very dangerous
Ion beam	Particle accelerator	Produced positively charged ions are accelerated at a high speed (around 20%80% of the speed of light) deposit high energy on a target	Dangerous

However, it can be used for irradiating whole plants and delicate materials, such as pollen grains. Various mutants have been developed through gamma radiation. The mutagenic effect results mostly from DNA double-strand breaks. The mutants show higher potential for improving plant architecture leading to better crop improvement and are used as a complementary tool in plant breeding. Gamma rays have a shorter wave length and therefore, possess more energy than protons and X-rays, which gives them ability to penetrate deeper into the tissue. Neutrons are hazardous and hence have less penetrating abilities, but they are known to cause serious damage to the chromosomes. They are best used for materials, such as dry seeds. Various forms of neutrons were also studied extensively for their use in mutagenesis in the 1960s and 1970s. Though it has been proved to be an effective mutagen, particularly for producing large DNA fragment deletions, the application of neutrons in induced mutagenesis is limited. The mutagenic effect of ultraviolet light was discovered by Altenbung through irradiation of the polar cap cells of fruit fly eggs. The mutagenic potential of these rays have since been confirmed in many organisms. In those organisms, germ tissue could be easily exposed to the low-penetrating ultraviolet light which resulted in covalent dimerization of adjacent pyrimidine. Emission of UV light (250290 nm) has a modest capacity to infiltrate tissues as compared with ionizing radiation. Ionizing emission goes deeper into the tissue and can cause a great number of variations in the chemical composition. The major advantage of using physical mutagenesis compared to chemical mutagenesis is the degree of accuracy and sufficient reproducibility, particularly for gamma rays, which have a uniform penetrating power in the tissue. During the past two decades, ion beams either through implantation or irradiation have become a new type of physical mutagen instead of the widely used gamma rays, X-rays and neutrons. They consist of particles travelling along a path that vary in mass from a simple proton to a uranium atom which are generated through particle accelerators. The positively charged ions are accelerated at a high speed (about 20%80% of the speed of light) and form high linear energy transfer (LET) radiation. LET radiation causes significant biological effects, such as chromosomal aberration, lethality, etc., as compared to other types of radiation used in physical mutagenesis. The damage caused by ion beams to DNA double strands is less repairable when compared to that induced by gamma rays due to deletion of DNA fragments of various sizes. More recently, to study the intricacies of mutation induction in space, plant materials have been sent out into aerospace. It has been speculated that the special environment of space flight, such as cosmic radiation, microgravity, weak geomagnetic field, etc. contains the potential agents of mutation induction. However, knowledge of the underlying genetics of aerospace mutagenesis is so far scarce.

From the past to the present, doses that lead to 50% lethality (LD50) have often been chosen. As noted by Oldach, it can be argued that LD50 is quite arbitrary and might lead to a high number of (mostly deleterious) mutations in every plant. This could go to the extent that desirable mutations are either lost or overlooked due to either plant mortality or poor agronomic performance in generations following the mutagenesis. Therefore, a mutation rate targeting a lower LD (e.g. LD20) with a survival rate of 80% appears to be more suitable for mutation breeding in selfing plant species. Maluszynski et al. also suggested that the final doses for mutagenic treatment should be rather low if the aim is to add new traits to an already high-quality genetic background, such as varieties or elite breeding lines. They conclude that the doses with an LD50 generally applied in the mutation breeding programmes of the 1960s and 1970s were too high and thus did not lead to the success expected from this technology. Ionizing radiations cause mutations by breaking chemical bonds in the DNA molecule, deleting a nucleotide, or substituting it with a new one. Acquaah also pointed out the importance of radiation being applied at the proper dose, a factor that depends on radiation intensity and duration of exposure. The dosage of radiation is commonly measured in roentgen (r or R) units. The exposure may be chronic (continuous low dose administered for a long period) or acute (high dose over a short period). The quality of mutation (proportion of useful mutations) is not necessarily positively correlated with dose rate. It is common knowledge that a high dose does not necessarily yield the best results. The mutagen dose used should be a compromise between mutation load and the chance to find desirable mutations, and this greatly depends on the cost effectiveness of selection. Screening of larger mutant populations that originate from a lower mutagen dose may be feasible for traits with simple phenotypic selection criteria, such as early maturity. On the other hand, screening the same population for complex phenotypic traits, such as seed protein quality would not be feasible.

Table: Examples of commonly used chemical mutagens.

Mutagen group	Example	Mode of action
Alkylating agents	1-methyl-1-nitrosourea (MNU); 1-ethyl-1-nitrosourea (ENU); methyl methanesulphonate (MMS); ethyl methanesulphonate (EMS); di-methyl sulphate (DMS); diethyl sulphate (DES); 1-methyl-2-ni-tro-1- nitrosoguanidine (MNNG); 1-ethyl-2-nitro-1- nitrosoguanidine (ENNG); N,N-dimethylnitrous amide (NDMA); N,N-diethylnitrous amide (NDEA)	React with bases and add methyl or ethyl groups and, depending on the affected atom, the alkylated base may then degrade to yield an abasic site, which is mutagenic and recom-binogenic, or mispair to result in mutations upon DNA replication.
Azide	Sodium azide	Same as alkylating agents.
Hydroxylamine	Hydroxylamine	Same as alkylating agents.
Hydroxylamine	Hydroxylamine	Same as alkylating agents.
Antibiotics	Actinomycin D; mitomycin C; aza-serine; streptonigrin	Chromosomal aberrations also reported to cause cytoplasmic male sterility.
Nitrous acid	Nitrous acid	Acts through deamination, the replacement of cytosine by uracil, which can pair with adenine and thus through subsequent cycles of replication lead to Transitions.
Acridines	Acridine orange	Intercalate between DNA bases thereby causing a distortion of the DNA double helix and the DNA polymerase in turn recognizes this stretch as an additional base and inserts an extra base opposite this stretched (intercalated) molecule. This results in frame shifts, i.e. an alteration of the reading frame.
Base analogues	5-bromouracil (5-BU); maleic hydrazide; 5- bromodeoxyuridine; 2-aminopurine (2AP)	Incorporate into DNA in place of the normal bases during DNA repli-cation thereby causing transitions (purine to purine or pyrimidine to pyrimidine); and tautomerization (existing in two forms which inter-convert into each other, e.g. guanine can exist in keto or enol forms).

Chemical Mutagenesis

The effect of chemical mutagens on plant materials is generally considered milder. An advantage of chemical mutagenic agents is that they can be applied without complicated equipment or facilities. The ratio of mutational to undesirable modifications is generally higher for chemical mutagens than for physical mutagens. Usually, the material is soaked in a solution of the mutagen to induce mutations. However, chemical mutagens are generally carcinogenic and therefore, extra care must be taken for health protection during the procedure. Material and safety data sheets for the specific chemical mutagen chosen should be carefully read and the agent should be appropriately inactivated before disposal. Despite the large number of mutagenic compounds, only a small number has been tested in plants. Among them, only a very restricted group of alkylating agents has found large application in plant experimental mutagenesis and plant mutation breeding. Over 80% of the registered new mutant plant varieties reported in the International Atomic Energy Association (IAEA) database obtained via chemical mutagenesis were induced by alkylating agents. Of these, three compounds are significant: ethyl methane sulphonate (EMS), 1-methyl-1-nitrosourea and 1-ethyl-1- nitrosourea, which account for 64% of these varieties.

Alkylating agents can be found among a large array of classes of compounds, including sulphur mustards, nitrogen mustards, epoxides, ethyleneimines, ethyleneimides, alkyl methanesulphonates, alkylnitrosoureas, alkylnitrosoamines, alkylnitrosoamides, alkyl halides, alkyl sulphates, alkyl phosphates, chloroethyl sulphides, chloroethylamines, diazoalkanes, etc. One of the most effective chemical mutagenic groups is the group of alkylating agents (these react with the DNA by alkylating the phosphate groups as well as the purines and pyrimidines). Another group is that of the base analogues (they are closely related to the DNA bases and can be wrongly incorporated during replication). Examples are 5-bromouracil and maleic hydrazide. A clear advantage of the point mutations created by chemical mutagens is their potential to generate not only loss-of-function but also gain-of-function phenotypes if the mutation leads to a modified protein activity or affinity, like tolerance to the herbicide glyphosate or sulphonylurea shown in the legume Medicago truncatula. The concentration of the mutagen, the length of treatment and the temperature at which the experiment is carried out affect the efficiency of mutagenesis. As chemical mutagens are very reactive, it is important to use fresh batches of the chemical(s) that have been appropriately stored.

Types of Mutations

Mutations can be broadly divided into intragenic or point mutations (occurring within a gene in the DNA sequence); intergenic or structural mutations within chromosomes (inversions, translocations, duplications and deletions) and mutations leading to changes in the chromosome number (polyploidy, aneuploidy and haploidy). In addition, it is important to distinguish between nuclear and extranuclear or plasmone (mainly chloroplast and mitochondrial) mutations, which are of considerable interest to agriculture.

Among the various kinds of mutational changes at the molecular level are base substitutions, a term meaning nucleotide changes that involve substitution of one base for another. This can happen through mis-pairing of the base analogue in the treated DNA during replication, leading to mutation through transitions when exchanges occur either between purines (AG) or between pyrimidines (TC) and transversions when purines are exchanged for pyrimidines or vice versa (A, GT, C).

Basically, transitions and transversions are the simplest kinds of base pair changes, but they may result in phenotypically visible mutations. There are no restrictions on the different kinds of sequence changes in the DNA of a gene following different types of misprints during replication. Another common error would be addition or deletion of a nucleotide base pair when one of the bases manages to pair with two bases or fails to pair at all. These kinds of sequence changes resulting in an alteration in the reading frame of the gene's DNA are known as frame-shift mutations. They are more drastic in their effect as they may completely change the message of the gene starting with the point of deletion/addition. Some of the mutations occur from rearrangement of bases in the DNA. A small or large sequence of bases may be inverted as a result of chromosome breakage, and reunion of the broken ends may involve different DNA molecules in a reciprocal rearrangement or in loss of a fragment. Duplication of a DNA sequence is another common mechanism for change in the structure of a gene leading to gene mutation.

Practical Considerations in Induced Crop Mutagenesis

The dose of a mutagen that achieves the optimum mutation frequency with the least possible unintended damage is regarded as the optimal dose for induced mutagenesis. For physical mutagens, this is estimated by carrying out tests of radio sensitivity (from radiation sensitivity), a term described as a relative measure that gives an indication of the quantity of recognizable effects of radiation exposure on the irradiated subject. Its predictive value, therefore, guides the researcher in the choice of optimal exposure dosage depending on the plant materials and the desired outcome. Mba et al. further describe the steps and procedures used for induction of mutations in vegetatively propagated plants and seeds, based on established protocols validated for cassava and rice, respectively. These protocols are for chemical mutagenesis (using EMS) and physical mutagenesis (using gamma rays) and include a list of the required equipment and reagents. The protocols cover the procedures ranging from pre-treatment to mutagenic treatments to post-treatment handling of the propagules materials, as well as the methodologies for data collection and analysis.

Important factors influencing the outcome of mutagenesis using chemical mutagens include the condition of the mutagenic solution; inherent characteristics of the targeted tissue; the environment; concentration of mutagen; treatment volume; treatment duration; temperature; presoaking of seeds; pH (7.0); catalytic agents (Cu^{2+} and Zn^{2+}) and post-treatment handling. Factors influencing the outcome of mutagenesis using physical mutagens includes oxygen; moisture content; temperature; physical ionizing agents (EM and ionizing radiation); dust and fibres (e.g. from asbestos); biological and infectious agents (both viral and bacterial). In general, the steps required for inducing and detecting mutations vary among sexually and asexually propagated plants/ crops but there are some basic principles that they share in common.

- A perfect understanding of the genetic makeup of the traits to be improved is very important. For example, a trait controlled by many genes (i.e., polygenic) has less chances of inducing modification compared to a trait that is governed by a single gene (i.e., monogenic).
- Understanding the mode of reproduction of the target crop is also a prerequisite, whether asexually or sexually propagated. If it is asexually propagated, then the method to employ is the next question: whether it will be in vitro or in vivo. If the crop is seed propagated, the question will be on the type of fertilization (self or cross-fertilization) to be used.
- The determination of the material that is to be used for the propagation prior to treatment, i.e. gametes or seeds for sexually propagated crops; and stem cuttings, buds, nodal segments or twigs for asexually propagated ones.
- Knowledge of the number of sets of chromosomes in the nucleus of a cell (ploidy) of the

target crop, especially when it relates to how hybridization barriers could impact on the predicted effectiveness of the induced mutants.

- Determinations of the genetic pedigree of the target crop for inducing mutations, i.e. selecting homozygous plants and the best genotype that is deficient in a single trait.
- Selection of an appropriate mutagen (physical or chemical mutagens) and dose (duration and concentration of mutagens). That is why a pilot assay is advisable to be carried out prior to the large-scale treatment of propagules.
- Identification of infrastructure (irradiation house, laboratories, screen/glass house, fields, etc.) for successful selection of desired mutants.
- Screening techniques for dissociation of chimeras from stable mutants.

Mutation Breeding Strategy for Obtaining Mutants

Any mutation breeding strategy requires several sequential steps. The effectiveness of mutation breeding over other breeding methods depends upon the efficacy of selection of useful variant mutants in the second (M2) or third (M3) generation. The first step in mutation breeding is to reduce the number of potential variants among the mutagenized seeds or other propagules of the first (M1) plant generation to a significant level to allow close evaluation and analysis. Determination of the target population size in the first generation of mutants is a prerequisite for potential success in any mutation breeding programme. The targeted population should be fixed so as to allow a high number of mutation measurements.

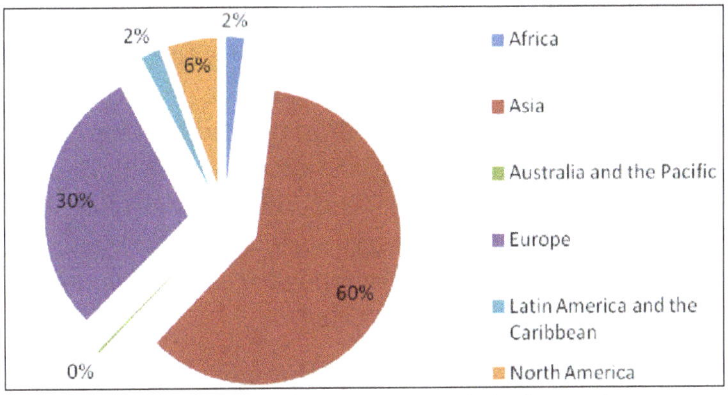

Figure: Distribution of mutant crop varieties by continents.

Thus, the population size should be managed effectively by the breeder. It should be noted that the population size depends on the inheritance pattern of the target gene. Therefore, it is advisable to select mutagens that give a high mutation frequency so as to reduce the population size of the M1 generation. Genetically, Ml mutant plants are heterozygous. This is because only one allele is affected by one mutation during treatment. However, the probability of having a mutation on both the alleles concurrently is a product of individual probability of mutation. Therefore, its occurrence is extremely low. Moreover, in Ml, only dominant mutations can be identified, while it is impossible to identify a recessive mutation expression at this stage. In this case, a plant breeder should attempt screening mutations in subsequent generations where segregation will occur. Consequently, the plant breeder generates homozygotes for dominant or recessive alleles. Caution should be taken to prevent cross pollination among the M1 population as this would lead to generation of new variation which will be difficult to differentiate from the effect of mutation. Screening and selection start in the M2 generation. Roychowdhury et al. discuss three main types of screening/selection techniques. These are physical/mechanical, visual/phenotypic and other methods. Physical or mechanical selection can be used efficiently to determine the shape, size, weight, density of seeds, etc., using appropriate sieving machinery. Visual screening is the most effective and efficient method for identifying mutant phenotypes. Visual/phenotypic selection is often used in selection for plant height, adaptation to soil, growing period, disease resistance, colour changes, earliness in maturity, ion-shattering, climate adaptation, etc. In the category of 'others', physiological, biochemical, chemical, physio-chemical procedures for screening may be used for selection of certain types of mutants. When a mutant line appears to possess a promising character, the next stage is seed multiplication for extensive field trials. The methods for comparative trials of mutants are the same as those for any other newly developed varieties. The purpose of field trials is to find whether the mutant promises to become a commercial variety that is superior to the mother cultivar. Prior to release as a commercial variety, the promising mutant should be studied for combinations of different characters like growth habit, structure and yield components in a wide range of environments under varying water availability, plant density, sowing dates, etc.

Impact of Mutant Cultivars

Genetic variability as a result of induced mutation by various mutagens has contributed to modern plant breeding. Over the past five decades, it has played a major role all over the world in the development of superior plant varieties with characteristics of high-yield, early maturity, lodging resistance among others. Several achievements in crop improvement through mutation breeding have resulted in two major outcomes: improved varieties that are directly used for commercial cultivation and new genetic stocks with improved characters or with better combining ability of traits. These traits could be increased yield, enhanced nutritional quality, resistance to pest and disease, early maturity, drought and salt tolerance, etc. Although the development of new cultivars has been the primary objective of mutation breeding, the genetic stocks developed can have numerous applications in plant breeding, from being used as a donor parent in conventional breeding programmes or as a parent in hybrid breeding programmes. The technique of identification of a gene by knockdown of the phenotypic expression through induced mutagenesis is a major component of research on molecular genetics and genomics today.

- Direct use of a mutant line that is developed through physical and chemical mutagenesis, or somaclonal variation.
- Indirect use of a mutant line/lines, which is/are used as a parental variety/varieties in cross breeding (cross between mutant lines or with a commercial variety/varieties).
- Use of mutant gene allele (trait), e.g. the Calrose 76 sd1 allele (semi-dwarf 1 trait) in rice.
- Use of wild species' genes translocated into plant genomes through irradiation-facilitated translocations, e.g. genes of wheat wild relative species.

Mutagenesis in Rice

Rice, among all crops, has displayed the highest advances in functional genomics in recent decades.

It is a diploid species, with a small genome in comparison to other cultivated cereals. In addition, the wide use of genetic transformation techniques, the synteny with other crop species, and a diversified source of related and closely related germplasm contribute enormously to the advantages in using rice as a genetic system for functional analyses. However, elite rice cultivars exhibit narrow genetic variability due to the repeated use of similar genotypes, i.e., with the same ideotype in crossing blocks. Understanding the genetic basis and the function of a given gene can help breeders to develop new, more productive and stress tolerant cultivars. One has to keep in mind that most of the agronomically important traits are of complex inheritance and therefore more difficult to improve. In this case, the mutant or variant allele can be detected and easily introgressed by performing Genome Wide Association Studies (GWAS) in populations including mutant genotypes.

Mutations can be used as a tool for gene functional studies and to create genetic variability. The analysis of mutants by forward genetics, i.e., from phenotype to gene, or by reverse genetics, i.e., from gene to phenotype, can be used to understand gene function. However, spontaneous mutation rates in higher plants are low, ranging from 10–5 to 10–8. Thus, mutagenesis is an important strategy to increase mutation frequency enabling studies of functional genomics and the development of new genotypes.

Mutations can be randomly induced by genotoxic agents or by DNA insertions and both can lead to gene gain or loss of function. Another form is by inducing direct mutations, in order to change the structure of a target gene. In rice, some important mutant collections created with the aim of functional studies, have been reported and a mutant library list was provided by Hirochika et al. Within the library, one can find mutant collections for randomly induced mutations and insertional mutations by transposons, retrotransposon, T-DNA insertions and even Arabidopsis mutant lines expressing rice full-length cDNA. The mutants developed can help to define a pathway or be directly used in agriculture, a classic example being the imidazolinone herbicide-resistant rice cultivars.

Random Mutations

Mutations are the major source of genetic variability and artificial mutations can be induced by mutagens. There are three major tools for mutagenesis: biological agents such as transposons, retrotransposons, and T-DNA; physical agents such as ionizing radiations or chemical agents such as alkylating agents; and azides. The Food and Agriculture Organization of the United Nations/International Atomic Energy Agency – Mutant Variety Database data (2019) reports on the developed and officially released mutants, a total of 3,275 accessions from 225 species. The released mutants are mainly composed of cereal species (47%) with the majority (25%) being rice mutants. The 823 mutant rice events are distributed in 30 Countries, with China and Japan responsible for 35.6% and 26.8% of this total, respectively.

In rice, changes in the gene structure can be randomly caused by rupturing the DNA through physical or chemical agents. The mutagenesis in rice is advantageous due to its small genome, i.e., a small population is required to saturate the whole genome and to provide a larger allelic series for use in mutagenesis. Random mutations caused by physical and chemical agents have been applied to create genetic variability and for gene functional studies in rice.

Indeed, the degree of mutation depends on the tissue, dosage and time of exposure. An important step when obtaining a mutant population, either by chemical or by physical agents, is the LD50 determination. The highest mutation frequency is expected to occur in the treatment that kills 50% of the treated material. In this sense, the LD50 corresponds to the dose that kills half of the treated population. In rice, mutants were developed from immature embryos, calli obtained from seeds and cell suspension cultures. Seeds are, however, easier to handle and do not require a specialized structure, and is therefore, the most widely used material. When seeds are mutagenized, it is important to standardize the seed mutagenic absorption through the pre-soaking of the seed in distilled water, since it also activates the seed metabolism and helps the mutagen action.

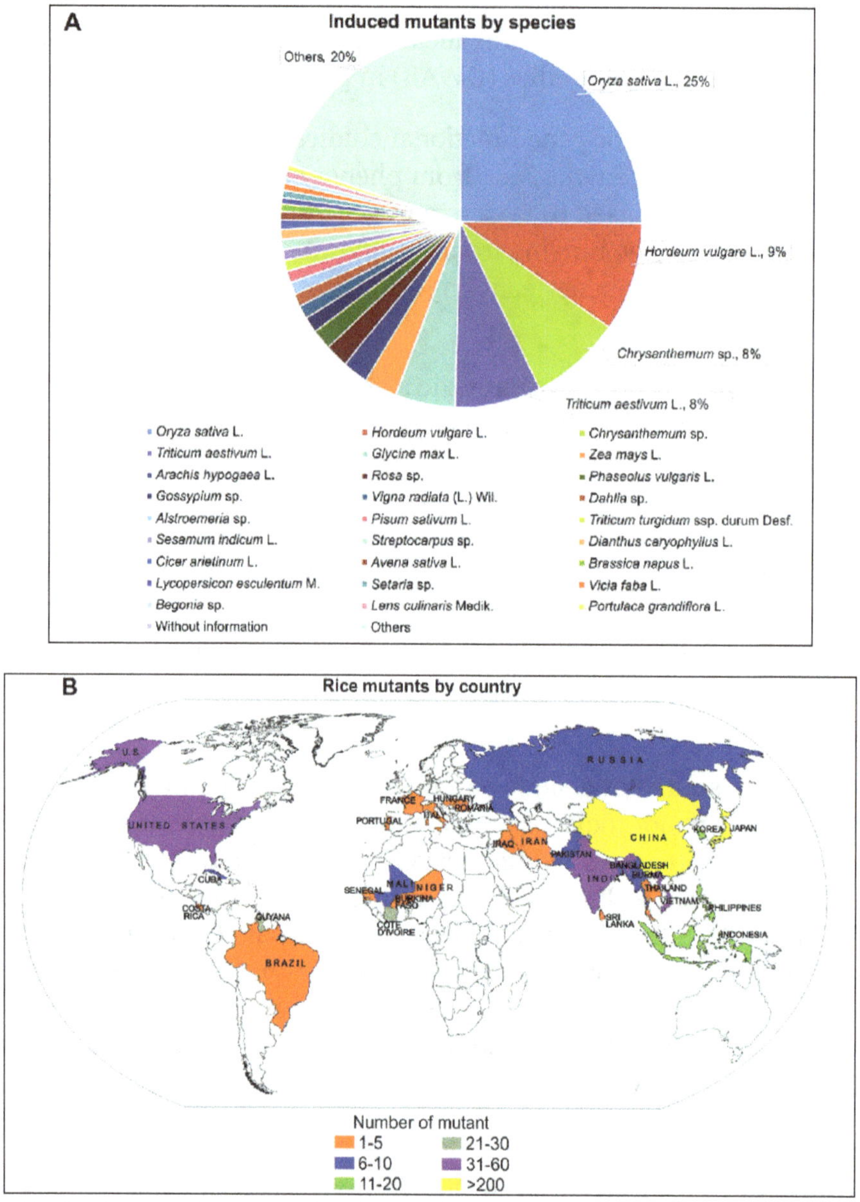

Figure: Mutant distribution according to Food and Agriculture Organization of the United Nations/International Atomic Energy Agency – Mutant Variety Database. (A) Distribution of mutants by species; (B) Distribution of mutants by country.

In plant breeding, the induction and identification of seed mutations is a simple process. Once the mutant population is obtained and the mutation is identified, seeds from the mutant plant compose the next generation in which the phenotype will be analyzed in order to search for mutation effects.

Mutagenesis is performed with a chemical or physical seed treatment given to an Mo population. The M1 plants originate from each treated Mo seed. Seeds from M1 plants will form the M2 population in which DNA analyses will be performed to find mutations. Seeds from M2 plants will form the M3 population and the next populations will be obtained in the same way.

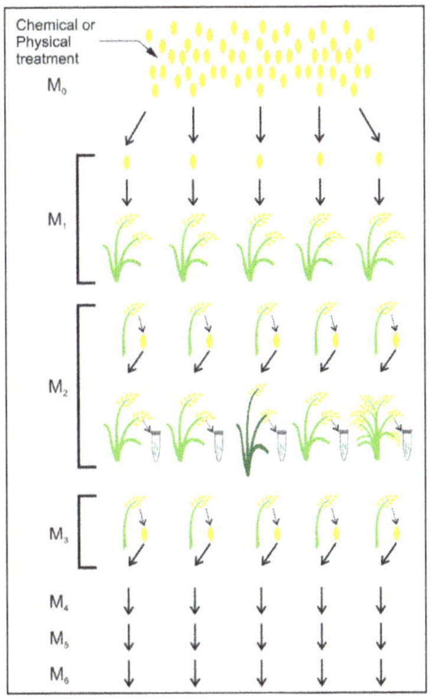

Figure: Scheme for obtaining a mutant population in rice.

Random Mutations Induced with Chemical Agents

The advantage of using chemical agents is the possibility to improve one or two traits while avoiding undesirable changes. The chemical mutagens ethyl methanesulfonate (EMS), methyl methanesulfonate (MMS), hydrogen fluoride (HF), sodium azide (SA), N-methyl-N-nitrosourea (MNU), and hydroxylamine (H3NO) are the chemicals most frequently used in plants. The effects of chemical treatment are silent or missense mutations (50%) while only 5% of nonsense mutations are observed.

Ethyl Methanesulfonate

Ethyl methanesulfonate (EMS) causes mutations through the alkylation of guanine bases leading to mismatches with thymine rather than cytosine, resulting in transitions of G/C to A/T. However, with less frequency, EMS can cause G/C to C/G or G/C to T/A transversions through 7-ethylguanine hydrolysis or A/T to G/C transitions through mismatches of 3-ethyladenine. In rice, EMS mutagenesis involves seed soaking in a solution with a known EMS concentration, usually 0.2 to 2.0%. The time in the soaking solution ranges from 10 to 20 h and considers the sensibility or the lethal curve of the genotype used. Luz and co-workers generated 340 rice mutant families applying EMS 1.5% for 2 h and a significant difference between the families was reported, indicating the efficiency of the dose and the time of exposition in generating mutants. EMS treatment was also reported to induce genetic variability in Basmati rice. For EMS treatment, presoaked seeds were immersed in EMS 0.5, 1.0, 1.5 and 2.0% for 6 h, suggesting that the 0.5% to 1.0% doses for 6 h, can induce mutations in rice. Since EMS mutagenesis causes a higher number of non-lethal point mutations, a relatively small population, ca. 10,000 plants, is necessary to saturate the genome with mutations. In diploid organisms, EMS treatment (1.5%) can induce 1 mutation every 294 Kb.

Reports have demonstrated that the application of EMS improves agronomically important traits. One of the focuses in rice mutagenesis is to obtain a genotype with C4 photosynthesis. The mutagenesis with EMS allowed one to obtain mutants affecting important traits such as increased vein number and the photosynthetic rate in leaves and reduced mesophyll interveinal cells. Traits involved in abiotic stress tolerance were also obtained. Among these, traits for drought tolerance such as root length, the volume of roots, and root weight were increased. Heat and drought tolerance were associated to low ROS accumulation, which is related to these stresses, and mutants showing higher chlorophyll retention were also heat tolerant. Furthermore, a heat tolerant mutant with higher photosystem II efficiency was identified. These studies demonstrated the potential use of the mutation for higher yielding, stress resilient genotype development. On the other hand, EMS mutagenesis was used in reverse genetics studies to identify gene function. The OsHAC4 gene that codes for rhodanase-like protein was identified and reported to be involved in arsenate tolerance mechanism and in arsenic accumulation in rice.

N-Methyl-N-Nitrosourea

Undoubtedly, N-methyl-N-nitrosourea (MNU) is the most applied mutagen in rice. MNU is a monofunctional alkylating agent and presents high reactivity with oxygen atoms in the DNA molecule. Mutation initiates with the addition of alkyl groups to nitrogenous bases, mainly guanine and cytosine. The generation of O 6-alkylguanine, induced by MNU treatment, promotes G/C to A/T transitions through mismatches with thymine during DNA replication. However, O 6-alkylguanine occurrence is not random and is influenced by the guanine 5' and 3' flanking sequence, DNA conformation and chromatin structure. If the 5'guanine is flanked by a 5'adenine and 3'cytosine, it shows a 10-fold increase in chance to form O 6-alkylguanine. It was reported that mutagenesis with MNU in rice is more efficient in developing cells than in dry seeds. When treatment with MNU is applied, one mutation every 135 Kb is expected.

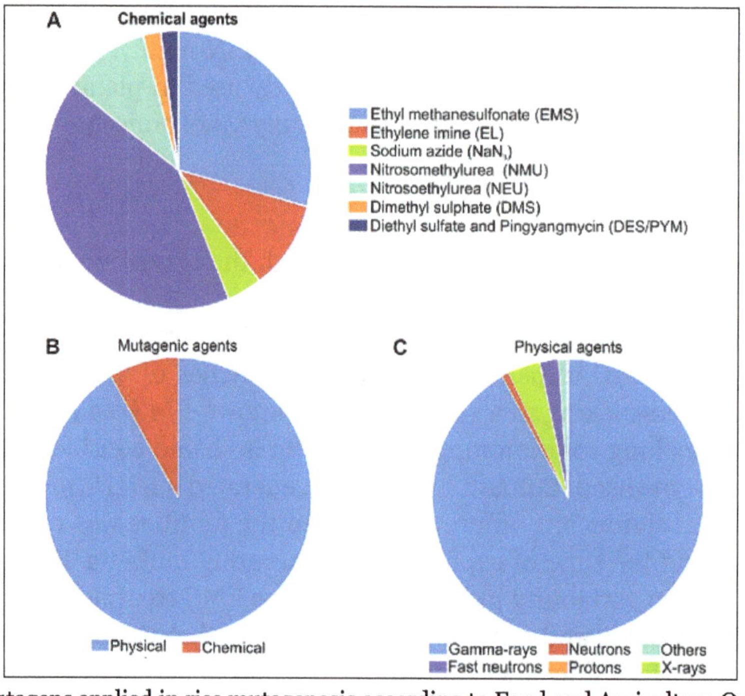

Figure: Mutagens applied in rice mutagenesis according to Food and Agriculture Organization of the United Nations/International Atomic Energy Agency – Mutant Variety Database. (A) Chemical mutagens; (B) Chemical and physical agents; (C) Physical mutagens.

Mutagenesis with MNU treatment has been widely applied and affects a range of plant physiological processes leading to the discovery of gene functions and the increase of genetic variability. With molecular and phenotypic markers, the locus stg(t) (responsible for the Stay Green trait) was mapped on chromosome 9. Spotted Leaf Mutants (spl28) were identified and displayed the formation of hypersensitive lesions leading to the initiation of leaf senescence. The gene Young Seedling Stripe2 (YSS2) was also identified and characterized to be involved in chloroplast biosynthesis. Additionally, the Plastochron1 (Pla1) mutant, which is involved in leaf primordia development and affects the timing of leaf vegetative growth, was also identified. The LGF1 (Leaf Gas Film 1) gene was found to control leaf hydrophobicity and the formation of gas films.

The silicon efflux transporter gene Low Silicon Rice 2 (Lsi2) was also identified through MNU mutagenesis. Also related to transport, the Zebra3-1 (z3-1) gene was characterized to be involved on citrate transport and distribution during leaf development. In addition, mutant lines with higher cadmium accumulation showed low OsHMA3 expression that suggested the involvement of this gene in cadmium accumulation. Genes involved in auxin efflux and signaling were identified. The Ospin-Formed2 (Ospin2) gene, that codes for auxin-efflux carrier proteins and Giant Embryo (GE) that controls the expression of auxin and cyclin responsive genes, were identified.

Findings using MNU also help to understand rice responses against biotic stresses. The Blast Lesion Mimic (BLM) was identified to confer resistance against fungus. A different mutant line, XM14, showed resistance to six races of Xanthomonas oryzae pv. Oryzae, in which the Xa42 gene, responsible for resistance responses, was identified.

In addition, through MNU mutagenesis, a better understanding of the embryogenic processes in rice was achieved. The Shootless2 (SHL2), SHL4/Shoot Organization2 (SHO2), and SHO1 genes were reported to be involved in normal shoot apical meristem (SAM) development. The Lonely Guy (LOG) was also identified, activating cytokinin and regulating SAM activity.

Mutation induction with MNU also focused on grain yield and quality traits, which undoubtedly are the most important agronomic traits. The Aberrant Panicle Organization1-1 (APO1-1) allele was identified involved in inflorescence architecture, floral organ identity and leaf production. Also, the Panicle Apical Abortion (PAA-Hwa) gene was reported to be involved in panicle identity. Through the introduction of the indica SSIIa and GBSSI genes in a BEIIb japonica mutant genotype, it was possible to generate lines displaying higher resistant starch and amylose contents. The Floury Endosperm8 (FLO8) gene was identified as a regulator of starch biosynthesis related genes. Regarding seed yield and shape, the Dense and Erect Panicle 2-3 (DEP2-3) was reported to play a role in rice seed size and shape. The rice grain oil content was modified by MNU treatment and mutant lines showing increased triacylglycerol content were also identified. The content of flavonoids, phenolic compounds and antioxidant activity was also increased in rice seeds. The studies described clearly illustrate the efficiency in the induction of mutation in rice treated with MNU and its potential application in rice functional genomics and improvement.

Sodium Azide

Sodium azide (SA- NaN3) is an ionic compound and its mutagenicity is mediated through an organic metabolite (analogous to L-azidoalanine) of the azide compound generated by O-acetylserine sulfhydrylase enzyme. This metabolite enters the cell nucleus and interacts with the DNA, creating point mutations in the genome. In rice, SA predominantly causes transitions from G/C to A/T and similar sequences at the mutation site were identified as 5'-GGR-3 '. The mutation rate identified was 1.4 to 2.9 mutations per Mb. It is known that the mutagenic effect of SA depends strongly on the solution pH and can be potentiated when germinated seeds are treated.

Important findings were obtained through SA treatment towards rice improvement. The Suweon 542 is mutant rice with potential application on rice flour production, showing floury grains and small and less degraded starch particles. In this mutant, the recessive locus flo7(t) explains 92.2% of the grain flour variation. SA treatment in red rice also increased the amounts of proto-cyanidins, γ-oryzanol, vitamin E and iron accumulation. Increased iron and zinc amounts in polished grains, starch and amylose contents and production of new aromatic rice were also observed. Abiotic stress tolerant lines were also obtained with SA treatment. The salt hypersensitive 1 (shs1) mutant was reported to play a role in antioxidant metabolism and Na+ homeostasis under salt stress. The tms8 gene was found to be involved in genetic male sterility induced by temperature.

Although less prominent than EMS and MNU, SA is an alternative tool for mutagenesis and can create variability for rice functional genomics and breeding, respectively.

Colchicine for Polyploid Induction

Polyploidization is an important event for plant evolution. When the ploidy is increased, the plant yield tends to increase, and sometimes organs tend to get bigger. This phenomenon suggests polyploidization as an improvement strategy towards the production of a better rice plant. Colchicine is a chemical mutagen generally used to induce chromosome doubling. It is a mitogen disrupting agent which affects the microtubules that organize chromosomes in one side of the cell promoting symmetric cell division. The induction of polyploidization in rice is generally performed through colchicine treatment in callus derived from anthers or seed cultures. In some cases, polyploidy induction causes genome rearrangements which can negatively affect seed production. After a long period of investigation, two polyploid rice lines with Polyploid Meiosis Stability (PMeS) genes were obtained from progenies of polyploid hybrids from indica-japonica crosses. The PMeS-I and PMeS-II lines showed different traits compared to other polyploid lines such as increased seed yield and meiosis stability.

With the aim of exploring heterosis in rice, two rice polyploid lines with photoperiod- and thermo-sensitive genic male sterility were developed. These lines showed high cross pollination ability, changes in fertility, and high combination ability. Hybrids obtained from these lines showed high heterosis and the potential to increase rice grain yield and quality. The transcriptome of rice neo-tetraploids revealed a group of genes associated with fertility and heterosis. Among them, transcription factors, methyltransferases, photosynthesis related genes, metabolic processes, transport, fertility, resistance, epigenetic elements, meiosis and retrotransposons were identified showing the complexity of regulatory mechanisms associated with fertility and heterosis . This demonstrates that polyploidy induction may be a way to potentialize heterosis in self-crossing plants such as rice.

To understand the regulatory mechanisms related to genome duplications in rice, different studies were performed. DNA methylation analyses in a rice autotetraploid revealed class II transposable element (TE) hypermethylation at CHG and CHH sites. The increase in TE methylation inhibits its transposition, stabilizing the chromosomal integrity, and can suppress the expression of neighboring genes, potentially decreasing the deleterious effects of the dosage of these genes.

Regarding abiotic stress responses, chromosomal duplication in rice resulted in increased tolerance to salinity. Also, autotetraploids presented more salinity stress tolerance during germination and reduced mortality rate during the initial growth. In addition, autotetraploid rice presented higher proline and soluble sugar contents, as well as decreasing malondialdehyde content when compared to diploid rice.

A synthetic allopolyploid rice (DSAR) technology was developed to produce an allopolyploid rice which involves hybridization with distant genotypes following chromosome doubling. DSAR technology comprises wild rice crossing followed by embryo rescue and chromosome doubling with colchicine treatment.

Random Mutations Caused by Physical Agents

Physical agents are the most used mutagens in rice, corresponding to 91.6% of the reports found. Ionizing radiation can be divided in two classes according to the linear energy transfer. The higher rates of linear energy transfer occur in alpha, neutrons and heavy ion beams while small rates are found in γ-rays, X-rays and electron beams. Ionizing radiation produces reactive oxygen species (ROS) which interact with the DNA causing oxidative damage, nucleotide changes and single strand or double strand breaks.

Gamma-Rays

Gamma-rays (γ-rays) have been widely used to generate mutants in rice, ca. 92% of the rice mutants obtained with physical agents, were generated with γ-rays. Small deletions (1–16 pb) were the most frequent effect of γ-rays on the rice genome, however larger deletions (9.4–129.7 kb) and large fragment inversions (1284.8 – 3208.5 kb) were also detected. Indels and single base substitution were also identified, with a higher frequency of heterozygous when compared to homozygous mutations. Also, the treatment of dry seeds resulted in heritable mutations at frequencies of 7.5×10−6 to 9.8×10−6. Reports have indicated that γ-rays generate genetic variability in a wide range of rice genotypes, being an effective tool for rice improvement.

Mutants induced with γ-rays help with the identification of genes related to specific physiological processes in rice plants. The necrotic lethality1 (nec1) gene, when heterozygous, controls leaf necrosis processes. Analyses in the early-senescence-leaf (esl) mutant showed the involvement of the NADPH oxidase in the abscisic acid (ABA) signaling through O_2 - radical production. Gamma-rays were able to create genetic variability for abiotic stress tolerance, such as salinity in ST-87 and ST-301 lines, as well as plant height, tiller number, shoot and root weight, total biomass and panicle length. The arsenic-tolerant type 1 (aatt1) showed tolerance to arsenic in the soil.

Changes in rice grain content were also reported. The OASA1 mutant showed increased tryptophan storage, which enabled researchers to identify that rice tryptophan accumulation is influenced by the OASA1 gene structure. The MRXII mutant displayed elevated tocopherol content which was related to SNPs created in the OsVTE2 promoter, which generate a MYB transcription factor binding site. Aromatic rice was also obtained by γ-rays. These lines showed increased 2-acetyl-1-pyrroline (2AP) and decreased γ-aminobutyric acid (GABA) content. GM077 and GM645 mutants showed lower and higher amylose content, respectively, indicating that γ-rays also influenced starch storage in rice grains.

DNA changes using γ-rays were also the target of a project aiming to turn rice into a C4 plant. The altered leaf morphology (alm) showed reduced leaf interveinal distances, which is determined by the mesophyll cell size, an important anatomical adaptation to obtain a Kranz anatomy, typical in C4 plants. Also, regarding cell structure, Brittle Culm 3 (BC3) was characterized to be involved in secondary cell wall biosynthesis. The gene OsCLD1/SRL1 (Semi-Rolled Leaf 1) was characterized in the cell wall synthesis, epidermal integrity and water homeostasis pathways. Zebra2 (z2-2) was identified associated with the tetra-cis-lycopene accumulation.

Ion Beam Radiation

Ion beam radiation (IBR) differs from γ-rays concerning the linear energy transfer, and produces important deletions, greater than 1Kbp. IBR effects were verified in Arabidopsis thaliana. It causes point mutations (most of them deletions) and intergenic rearrangements as deletions, inversions, translocations and insertions. The DL50 of Carbon ions was estimated in 30 Gy (220 MeV) for the genotype Nipponbare. After, doses up to 20 Gy (23-40 keV/μm) and 10 Gy (60-80 keV/μm) in Carbon and Neon accelerated ion radiation, respectively, were recommended for the genotype Nipponbare to produce plants with high survival rates, fertility and with high frequency of chlorophyll deficiency mutants. IBR induces mutations with high frequency at a relatively low dose and also induces a broad spectrum of phenotypes without affecting other plant traits being these an advantage in applying IBR for rice mutation.

Applying IBR, the role of some important genes was unveiled in rice. Mutations were obtained in rice irradiated with Carbon ions, i.e., the OsNramp5 gene was changed, leading to a decreased Cadmium content in grains. Nitrogen bombardment also causes variations in some important traits, such as plant height, leaf, and tegument and pericarp color. PKOS1, HyKOS1 (dwarf) and TKOS4 (tall) mutants were obtained through bombardment with the low energy Nitrogen ion. It enabled the identification of the genes OsSPY and 14-3-3, which act as negative regulator of gibberellins and repressor of the RSG (Repression of Shoot Growth), respectively.

Fast-Neutron Irradiation

Fast-neutron irradiation affects the DNA structure, damaging nitrogenous bases promoting standard breaks. Also, single nucleotide substitution, insertions and duplications were reported as an effect of fast-neutron irradiation. Recently, 1,504 rice mutants generated by fast-neutron radiation on the Kitaake genotype were sequenced. More than 91,000 mutations were identified affecting 58% of the rice genes, with an average of 61 mutations in each line. These changes included single nucleotide substitutions, deletions, insertions, inversions, translocations and tandem duplications.

Despite the important DNA damage induced by fast-neutron irradiation, few reports have applied this tool as a source of variation in rice. However, important traits for rice grain quality were obtained. Changes in amylose content were detected in some mutant lines, while the irradiation did not alter the amylopectin structure. Tolerance to iron toxicity was obtained, fast neutron irradiation induced changes in the FRO1 gene, responsible to increase tolerance and higher iron accumulation in grains. Seeds from a rice genotype overexpressing NH1 (NPR1 homologue 1) gene treated with fast neutron irradiation provided information regarding immune responses against bacterial infection by Xanthomonas oryzae pv. Oryzae (Xoo). The NH1ox line showed Xoo resistance and the loss of function mutation of SNL1 gene (Supressor of NH1-mediated Lesion) caused suppression of the resistance conferred by NH1. Lately, the NH1ox-54 line was irradiated with fast neutron and new genes related with SNL1 activity were identified. RNAi lines obtained for CRK6 and CRK10 (cysteine-rich receptor-like kinases) genes displayed the snl1 phenotype indicating that these genes are required for the immune responses mediated by NH1. Recently in the NH1ox-54 irradiated line, a mutant for suppressor of NH1- mediated immunity 1 (snim1) which suppresses the resistance conferred by NH1 was identified.

Fast-neutron beams are highly important for functional genomics; however, it has not yet been applied in rice breeding. The characterized genes can be used towards the development of new varieties, since genes involved in agronomically important traits such as disease resistance, abiotic stress tolerance and grain quality have been altered.

Random Mutations Induced by Other Chemical and Physical Agents

Cosmic-ray radiation has been applied to induce genetic variability in rice. Seeds irradiated with γ-, cosmic-rays or IBR have been obtained. The ROS accumulation, peroxidases, ascorbate peroxidases and dismutase superoxide activity were increased in the three treatments, however only γ- and cosmic rays were able to decrease the carotenoids and chlorophyll contents. In addition, cosmic-ray radiation was able to change the up regulated and down regulated gene profiles compared to other types of radiation.

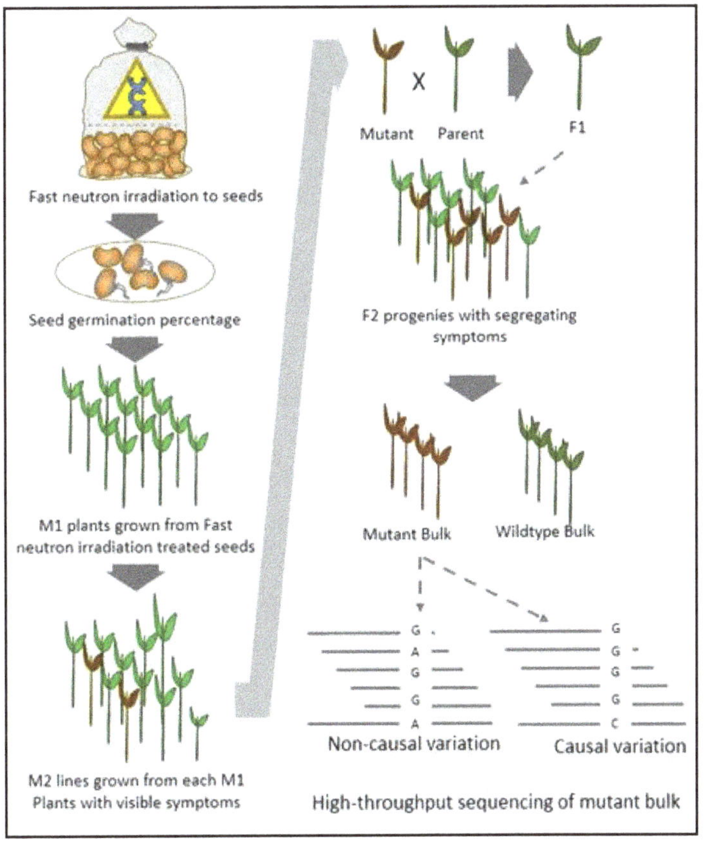

Diepoxybutane (DEB), a bifunctional alkylating agent, acts on nitrogen 7 of the guanines and nitrogen 3, 6, 7 and 9 of the adenines, promoting DNA-DNA and DNA-protein linkages. It is predicted that DEB cause small deletions (1Kb) and point mutation. Mutants developed with DEB treatment, showed variability for salinity tolerance. Tolerant mutants absorbed less Na^+ and more K^+, increased shoot biomass and elevated survival rate compared to the wild-type.

Mutation is an important tool in creating genetic variability. In this sense, there are several agents being tested that provide different effects on DNA. These studies are extremely important since they demonstrate the range of possibility by which the structure of DNA can be modified.

Random Mutations Induced by Combinations of Mutagens

Besides the chemical and physical mutagen's application described above, studies have demonstrated the combined effect of these agents in increasing the frequency of mutations. The combination of chemical and physical agents was reported in rice treated with γ-rays followed by EMS seed soaking. Salinity tolerant mutants were isolated and were able to survive up to 15 days in 342 mM NaCl solution while sensitive plants were already affected by 171 mM NaCl after 5 days.

The combination of γ-rays, EMS and SA was also used in Basmati rice mutagenesis. The combination of different doses promoted decreases in plant height, panicle size, and panicle number and in the weight of 100 seeds. Rice seeds irradiated with γ-rays, electron beam or recurrent treatment with γ-rays, showed that electron beam radiation presented higher mutation frequency for chlorophyll content compared with γ-rays and recurrent treatment.

References

- Molecular-markers-and-marker-assisted-breeding-in-plants, plant-breeding-from-laboratories-to-fields: intechopen.com, Retrieved 06, June 2020
- Principle-and-application-of-plant-mutagenesis-in-crop-improvement-A-review-287358946: researchgate.net, Retrieved 12, March 2020

Chapter 4

Techniques of Plant Breeding

Selection is an important process as it defines the characteristics of the offspring. Various selection methods are employed based on desired outcome and plant characteristics. Breeding is the next step and several breeding methods are available based on the plant's mode of reproduction and pollination. This chapter deals with the selection and breeding methods for self-pollinated, cross-pollinated, hybrid cultivars and clonally propagated plants in a comprehensive manner.

Breeding of Self-pollinated Plants

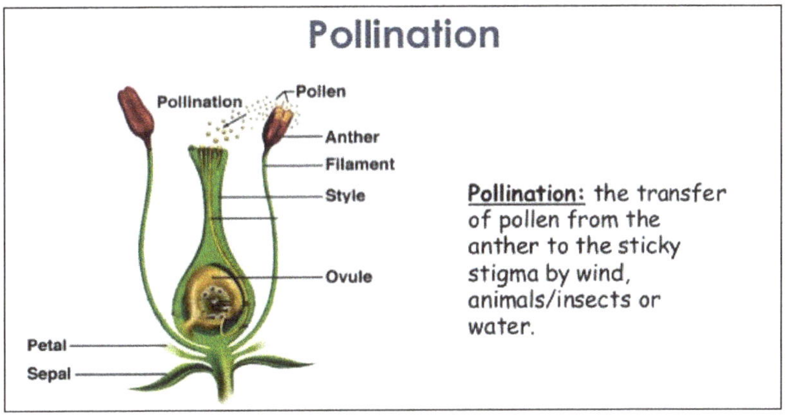

Self-pollination refers to a flower that has the ability to pollinate itself and other flowers produced by the same plant. It does not need another plant's pollen for fertility.

Hermaphrodites and monoecious species, which make up the majority of plant species, have the potential to self-pollinate.

A self-pollinating flower has both the male and female reproductive parts. Pollen will accumulate on the plant's male anthers and then transfer to the female stigma of the flower to complete the fertilization process.

Normally, plants that can self-pollinate feature distinctive carpels and stamens that are almost all the same length. Plants with the ability of self-pollination have the upper hand over plants that require cross pollination.

Self-pollinating plants can be pollinated by insects such as moths or bees, but they can also rely solely on the wind, airflow, or natural growth of the single flower to ensure that pollination occurs.

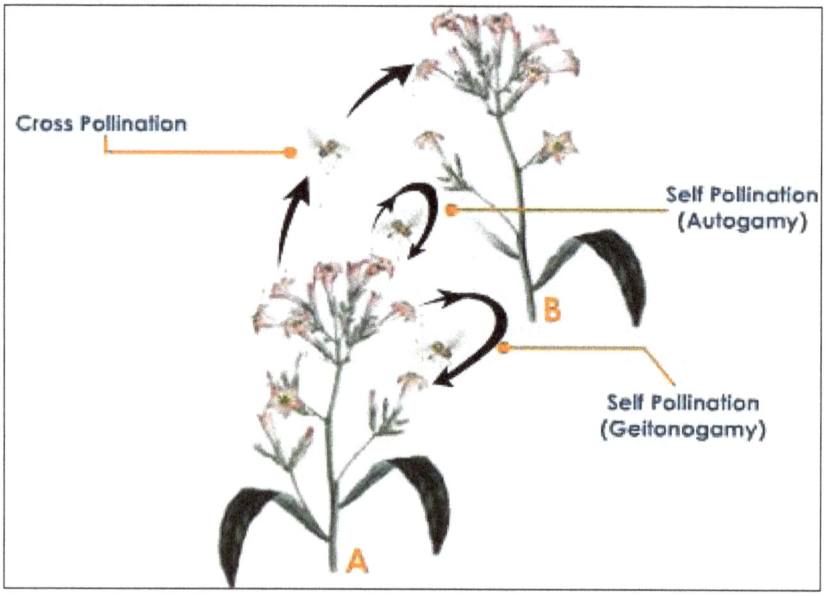

Some self-pollinating flowers can even be pollinated before they bloom by depending on the bud's natural growth. As the flower's stigma grows through the anthers and the flower begins to unfurl, it effectively transfers pollen between the flower's male and female parts.

Self-pollinated species have a genetic structure that has implication in the choice of methods for their improvement. They are naturally inbred and hence inbreeding to fix genes is one of the goals of a breeding program for self-pollinated species in which variability is generated by crossing. However, crossing does not precede some breeding methods for self-pollinated species.

The breeding methods that have proved successful with self-pollinated species are:

1. Mass selection;
2. Pure-line selection;
3. Hybridization, with the segregating generations handled by the pedigree method, the bulk method, or by the backcross method;
4. Development of hybrid varieties.

Mass Selection

Mass selection is often described as the oldest method of breeding self-pollinated plant species. However, this by no means makes the procedure outdated. As an ancient art, farmers saved seed from desirable plants for planting the next season's crop, a practice that is still common in the agriculture of many developing countries. This method of selection is applicable to both self-and cross-pollinated species.

Key Features

The purpose of mass selection is population improvement through increasing the gene frequencies of desirable genes. Selection is based on plant phenotype and one generation per cycle is needed. Mass selection is imposed once or multiple times (recurrent mass selection). The improvement is limited to the genetic variability that existed in the original populations (i.e., new variability is not generated during the breeding process). The goal in cultivar development by mass selection is to improve the average performance of the base population.

Applications

As a modern method of plant breeding, mass selection has several applications:

1. It may be used to maintain the purity of an existing cultivar that has become contaminated, or is segregating. The off-types are simply rogued out of the population, and the rest of the material bulked. Existing cultivars become contaminated over the years by natural processes (e.g., outcrossing, mutation) or by human error (e.g., inadvertent seed mixture during harvesting or processing stages of crop production).
2. It can also be used to develop a cultivar from a base population created by hybridization, using the procedure described next.
3. It may be used to preserve the identity of an established cultivar or soon-to-be-released new cultivar. The breeder selects several hundreds (200–300) of plants (or heads) and plants them in individual rows for comparison. Rows showing significant phenotypic differences from the other rows are discarded, while the remainder is bulked as breeder seed. Prior to bulking, sample plants or heads are taken from each row and kept for future use in reproducing the original cultivar.
4. When a new crop is introduced into a new production region, the breeder may adapt it to the new region by selecting for key factors needed for successful production (e.g., maturity). This, hence, becomes a way of improving the new cultivar for the new production region.
5. Mass selection can be used to breed horizontal (durable) disease resistance into a cultivar. The breeder applies low densities of disease inoculum (to stimulate moderate disease development) so that quantitative (minor gene effects) genetic effects (instead of major gene effects) can be assessed. This way, the cultivar is race-non-specific and moderately tolerant of disease. Further, crop yield is stable and the disease resistance is durable.
6. Some breeders use mass selection as part of their breeding program to rogue out undesirable plants, thereby reducing the materials advanced and saving time and reducing costs of breeding.

Procedure

The general procedure in mass selection is to rogue out off-types or plants with undesirable traits. This is called by some researchers, negative mass selection. The specific strategies for retaining representative individuals for the population vary according to species, traits of interest, or creativity of the breeder to find ways to facilitate the breeding program. Whereas rouging out and bulking appears to be the basic strategy of mass selection, some breeders may rather select and advance a large number of plants that are desirable and uniform for the traits of interest (positive mass selection). Where applicable, single pods from each plant may be picked and bulked for planting. For cereal species, the heads may be picked and bulked.

Steps

The breeder plants the heterogeneous population in the field, and looks for off-types to remove and discard.

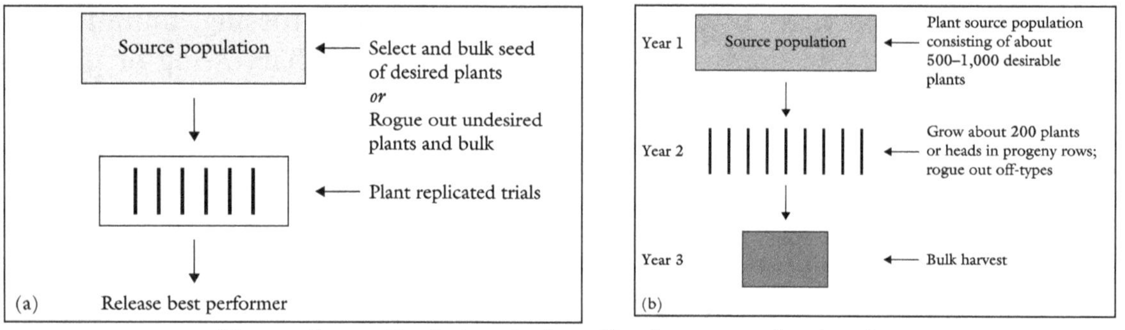

Figure: Generalized steps in breeding by mass selection for (a) cultivar development, and (b) purification of an existing cultivar.

This way, the original genetic structure is retained as much as possible. A mechanical device (e.g., using a sieve to determine which size of grain would be advanced) may be used, or selection may be purely on visual basis according to the breeder's visual evaluation. Further, selection may be based on targeted traits (direct selection) or indirectly by selecting a trait correlated with the trait to be improved.

- Year 1 If the objective is to purify an established cultivar, seed of selected plants may be progeny-rowed to confirm the purity of the selected plants prior to bulking. This would make a cycle of mass selection have a 2-year duration instead of 1 year. The original cultivar needs to be planted alongside for comparison.
- Year 2 Evaluate composite seed in a replicated trial, using the original cultivar as a check. This test may be conducted at different locations and over several years. The seed is bulk harvested.

Genetic Issues

Contamination from outcrossing may produce heterozygotes in the population.

Unfortunately, where a dominance effect is involved in the expression of the trait, heterozygotes are indistinguishable from homozygous dominant individuals. Including heterozygotes in a naturally selfing population will provide material for future segregations to produce new off-types. Mass selection is most effective if the expression of the trait of interest is conditioned by additive gene action.

In self-pollinated populations, the persistence of inbreeding will alter population gene frequencies by reducing heterozygosity from one generation to the next. However, in cross- pollinated populations, gene frequencies are expected to remain unchanged unless the selection of plants was biased enough to change the frequency of alleles that control the trait of interest.

Mass selection is based on plant phenotype. Consequently, it is most effective if the trait of interest has high heritability. Also, cultivars developed by mass selection tend to be phenotypically uniform for qualitative (simply inherited) traits that are readily selectable in a breeding program. This uniformity not withstanding, the cultivar could retain significant variability for quantitative traits. It is helpful if the selection environment is uniform. This will ensure that genetically superior plants are distinguishable from mediocre plants. When selecting for disease resistance, the method is more effective if the pathogen is uniformly present through- out the field without "hot spots".

Some studies have shown correlated response to selection in secondary traits as a result of mass selection. Such a response may be attributed to linkage or pleiotropy.

Advantages and Disadvantages

Some of the major advantages and disadvantages of mass selection for improving self-pollinated species are given here.

Advantages

1. It is rapid, simple, and straightforward. Large populations can be handled and one generation per cycle can be used.
2. It is inexpensive to conduct.
3. The cultivar is phenotypically fairly uniform even though it is a mixture of pure lines, hence making it genetically broad-based, adaptable, and stable.

Disadvantages

1. To be most effective, the traits of interest should have high heritability.
2. Because selection is based on phenotypic values, optimal selection is achieved if it is conducted in a uniform environment.

3. Phenotypic uniformity is less than in cultivars produced by pure-line selection.

4. With dominance, heterozygotes are indistinguishable from homozygous dominant genotypes. Without progeny testing, the selected heterozygotes will segregate in the next generation.

Modifications

Mass selection may be direct or indirect. Indirect selection will have high success if two traits result from pleiotropy or if the selected trait is a component of the trait targeted for improvement. For example, researchers improve seed protein or oil by selecting on the basis of density separation of the seed.

Pure Line Selection

Pure line selection is a method in which new variety is developed by selection of single best plant progeny among traditional varieties or land races.

Pure line is a self pollinated descendent of a self pollinated plant.

Desirable types already exist in population. Those are isolated through careful testing procedures. Another term used for this method of plant breeding is individual plant selection, as large numbers of plants are selected, but those are harvested individually, their individual progenies are grown and evaluated and then best progeny is released as a pureline variety.

Genetic Basis for Pureline Selection

Pureline varieties are homozygous and homogeneous as they are genetically similar and true breeding. Such varieties possess narrow genetic base so they are more susceptible to diseases, and have poor adaptability. A pure line breeding method is normally used for self-pollinated crops, but has importance in breeding of inbred lines that are used to make hybrids in self or cross pollinated crops.

Main steps in the production of pureline variety are as follows:

- Selection of individual plants from a local / traditional variety, land races or some other mixed population.
- Visual evaluation of individual plant progenies.
- Yield trials.
- Release of new pureline variety.

Detailed Procedure

Step number	Year	Details
One	First year	Many individual plants (about 200-3000) are selected from a local variety or some other mixed population based on phenotype. Any number of plants can be selected but importantly they should be homozygous. Seeds are harvested separately.
Two	Second year	Seeds from each plant are grown separately with proper spacing. Evaluated for characters under consideration. Probably in this first season it will not be possible to judge the value of separate lines well, but surely visually poor, weak and defective progenies can be rejected. Number of progenies is drastically reduced in this step, fewer are carried to step three.
Three	Third year	Seeds of second year crop are planted in a preliminary yield trial. Standard plots for comparison are introduced, undesirable progenies are rejected.
	Four to seven years	Replicated yield trials conducted at several locations. Standard plots for comparison are introduced, undesirable progenies are rejected. Tests are done for characters under consideration.
	Eighth year	Best progeny is released as a new pureline variety. Seed is multiplied for distribution.

Examples of Pureline

Crop	Variety	Features	Selection from and at
Cowpea	Suvita-2	Striga resistant line	Selection from landrace in Burkina Faso
Wheat	Kanred	Better in winter hardiness, rust resistance, earliness in maturity	Selected at Kansas from Crimean variety, an introduction from Russia
Oat	Columbia	Matures early, grain with high test weight.	Selected at Missouri station from Fulghum variety

Advantages of Pureline Selection

Merits of pureline selection are listed below:

- Easy and cheap method of crop improvement.
- Rapid method, lines are usually genetically fixed and yield trials can be immediately conducted.
- Plants in such variety react in similar fashion to environmental conditions, means they are uniform in performance and at the same time in appearance too.

- Maximum possible improvement over the original variety can be achieved.
- Useful in improving low heritability traits as selection is based on progeny performance.

Disadvantages of Pureline Selection

Demerits of pureline selection are given below:

- Pure lines have poor adaptability due to narrow genetic base, just opposite to mass selected variety.
- Superior genotypes can only be isolated from the mixed population. This selection is powerless to bring changes in hereditary factors i.e. to develop new genotype.
- Mostly popular or in fact limited to self pollinated spp. only.
- Time and space consuming.
- More expensive yield trials have to be conducted than in mass selection.

Applications of Pureline Selection

Applications of pureline selection are listed below:

- As pure line selection gives out uniformity in maturity, height etc. suitable for cultivars in which machines are used for the various production processes like harvesting.
- Because of the same reason as above it can be used for varieties grown for processing market, as uniformity in texture, canning qualities is important in here.
- It is practiced after hybridization within segregating populations.
- Though it can serve above purposes this method has limited practical use in the breeding of major cultivated species. However, the method is still widely used while breeding less important species that have not yet been heavily selected.
- Logically it was first step in the development of a uniform variety for the newly domesticated crops, with some amount of variability.

Hybridization

There are several methods of improvement of self-fertilized crops by hybridization. These are:

- Pedigree method or breeding.

- Bulk method or breeding.
- Single seed descent method.
- Back cross method.
- Multiple cross method.

Pedigree Method

Individual plant progeny is selected from F_2 and subsequent generations, and their progenies are tested. During this process the record of parents as well as off-springs is kept, for which it is known as pedigree method.

The pedigree is defined as the description of the ancestors of an individual and it is generally helpful in finding out the amount of relatedness among two individuals, i.e., whether they are related by common parent in their descent ancestry or not.

Procedure

First Year

The hybridization is done among two selected parents, after emasculation one become female parent and another male parent. After seed set and maturation, the F_1 seeds are harvested separately from each plant individually. On the basis of choice of parents, the type of cross will be of two types — it will be simple cross or complex cross.

Second Year

F_1 generation seeds are space planted and selfing is allowed, each F_1 will produce more F_2 seeds. From 15-30 selected F_1 plants, the F_2 seeds are collected to get a reasonable size of F_2 population and variation.

Third Year

In F_2 generation, 2000-10000 plants are space planted, 100-500 plants are selected and their seeds are harvested separately. If the parent plants are closely related varieties then the number of selected F_3 plants would be smaller whereas in case of distantly related varieties the number of F_3 progenies will be of relatively larger numbers.

Fourth Year

In F_3 generation also, the individual plant progenies are space planted. Each progeny should have about 30 or more plants. Individual plants with desirable characteristics are selected; disease and lodging susceptible progenies to be eliminated, and also the progenies with undesirable characters are rejected even from the selected plants.

During this selection if the number of superior progenies is very small then the whole cross programme may be rejected.

Fifth Year

The selection procedure is same as previous year, only if two or more progenies coming from the same F_3 progeny are similar and comparable, then only one may be saved and others may be rejected. The emphasis is given on the selection of desirable plants from superior progenies.

Sixth Year

Individual plant progenies of F_5 generation are planted according to recommended commercial seed rate. Three or more rows for each progeny will help in comparison among progenies. Many progenies may have become reasonably homozygous genotype and may be harvested in bulk. If the progenies show variation then the individual plants are selected. The number of selected progenies should be reasonable so that preliminary yield trial with 25-100 progenies can be done.

Seventh Year

Individual plant progenies of F_6 generation are planted in multi-row plots and evaluated visually. Progenies harvested in bulk since they become homozygous. The segregating progenies may be discarded and the preliminary yield trial may be done for the progenies which are reasonably homozygous and have enough seeds.

Eighth Year

Preliminary yield trial with three or more replications is conducted to identify few superior lines. The progenies are evaluated for plant height, lodging, disease resistance, flowering time, maturity time, etc. Quality test is done to serve as an additional basis for selection.

Ninth to Tenth or Thirteenth Year

The superior lines are tested in replicated yield trials at several locations. The above mentioned criteria are evaluated for these lines. The line which is superior to the best commercial variety may be released as new variety.

Eleventh or Fourteenth Year

The selected strain should get multiplied to release as a new variety. Breeder has the responsibility to supply the seeds to the state seeds corporation for production and marketing of the seeds.

Schematic Representation

First Year	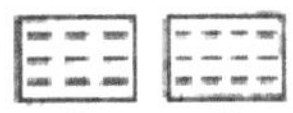 P_1 P_2	Selected plants of parental variety are hybridized to get the F1 hybrid seeds.
Second Year	 F_1	The hybrid seeds (10-30) are planted and the seeds of F2 harvesed n bulk.
Third Year	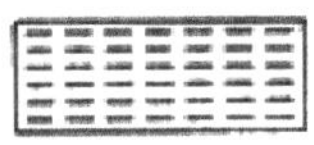 F_2	2000-1000 seeds are space planted. 100-500 superior plants are selected and seeds are harvested from those as pedigree
Fourth Year	 F_3	Individual plant progeries are planted in raw and space planted. Superior plants are selected
Fifth Year		The same procedure as in fourth year.
Sixth Year	 F_5	The individual selected plant progenies are in multi-row plots.
Seventh Year	 F_6	The same procedure as in previous year Enough seeds to be collected from superior plants.
Eight Year	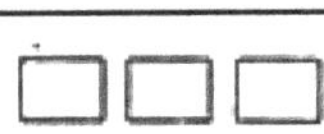 F_7	The Preliminary yield tiral and quality test has to be done.
Ninth to Thirteenth Year	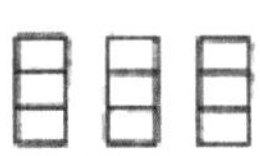 $F_8 - F_{12}$	Co-ordinated yield trials to be done. Disease and quality tests have to be done.
Fourteenth Year	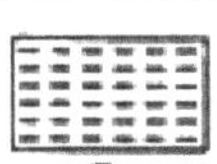 F_{13}	Multiplication of seeds for distribution and release as a new variety.

Merits

1. This method is most useful as transgressive segregation for yield and other quantitative characters may be recovered in addition to improvement of specific characters.

2. This method is well suited for improvement of characters, which can be easily identified and simply inherited.

3. Through the maintenance of pedigree record the breeder may be able to obtain the information about inheritance of characters.
4. Plants or progenies with weaker and visible defects are eliminated at an early stage in the breeding programme.
5. This method gives maximum importance on the breeder to use his/her skill and judgement about the selection of plants and progenies.
6. This method takes less time than bulk method to release a new variety.

Demerits

1. The success of the method is mainly dependent on the skill of the breeder.
2. To keep the individual pedigree record is laborious and time consuming, it may be the limiting factor for large breeding programme.
3. Selection of large number of progenies in every generation is also laborious and time consuming.
4. In F_2 and F_3, the selection for yield is not effective. If sufficient number of progenies is not retained, valuable genotypes may be lost in early segregating generations.

Achievements

Pedigree method is useful in selection of new superior recombinant types from a hybridization programme. This method is suitable for improving specific characteristics, such as disease resistance, plant height, maturity time, etc. as well as yield and quality characters.

Many improved varieties have been developed through pedigree method in many crops like wheat, rice, barley, pulses, oil seeds, cotton, tobacco, jowar, vegetables, etc.

Wheat

K 65 (tall variety) ← C 591 x NP 773.

K 68 (good quality grain) ← NP 773 x K 13.

WL 711 (dwarf, high yield) ← (S308 x Chris) x Kalyan Sona.

Malviya 12 (good grain) ← NP 876 x Cno 66.

Rice

'Jaya' and 'Padma' (short duration, finer grain) ←Taichung Native 1 x T 141.

Cotton

'Laxmi' (fibre quality, early maturing, resistant to leaf blight) Gadag 1 x CC 2 (Cambodia Coimbatore 2).

Tomato

Pusa early dwarf (more yield) ← Meerut x Red cloud.

Bulk Method

It is a method which can handle segregating generations, in which F_2 and subsequent generations are harvested in bulk to grow the next generation. At the end of bulking period, individual plant selection and evaluation is carried out in the similar fashion as in the pedigree method.

Procedure for Bulk Method

Bulk method of plant breeding completes in following main steps.

Step	Details
Hybridization	Crossing among selected parents.
F_1 generation	F_1 seeds (minimum 20) planted. Bulk harvesting is carried out.
F_2 to F_6 generation	F_2 to F_6 are planted, harvested in bulk. Number of plants should be as large as possible. Generally artificial selection is not carried out.
F_7 generation	Generally 30 to 50 thousand F_6 seeds are space planted, selection is carried out based on phenotype and 1000 to 5000 selected ones are harvested separately.
F_8 generation	Individual plant progenies are grown, inferior progenies eliminated. Harvested in bulk.
F_9 generation	Preliminary yield trials with standard varieties as check. Selection is based on yield.
F_{10} to F_{13} generation	Multi location yield trials are conducted using standard varieties as check. Evaluation of performance of lines is done.
F_{14} generation	Seed multiplication for distribution.

Advantages of Bulk Method

Here are the merits of bulk method.

- Simple and inexpensive.
- Little record keeping.

- Easy to handle populations as harvest in bulk.
- Natural selection is effective, possible to carry out artificial selection if required.
- We can wait for opportunity to carry out selection. Selection for some environment dependent characteristics like disease resistance, lodging resistance can be carried out when environment is favorable for disease epidemic and severe lodging. Till that year the bulk can be maintained easily.

Disadvantages of Bulk Method

Demerits of bulk method are given here:

- Requires more time duration to develop new variety.
- Pedigree record is not maintained, so we can not trace back progeny to the parent plant.
- Not suitable method for greenhouses.
- Large number of progenies has to be selected at the end of bulking period.
- Natural selection may also work against desirable traits.
- Not much scope for skills.
- Bulk method is particularly suited to small grain crop plants.
- Maximum productivity is established in F_2 generation of method.
- No recombination occurrence among superior lines.

Applications of Bulk Method

- Homozygous lines can be isolated in less time, automatic increase in homozygosity up to F_6 or F_8generation after this individual plant selection can be carried out.
- Natural selection may improve considered character (yield), with minimum expenses.
- Suitable for crops which are generally planted at high planting densities, e.g. small grain crops.

Single Seed Descent Method

This method was suggested by Coulden for advancing segregating generation of self-pollinated crops. A breeding procedure used with segregating populations of self-pollinated species in which plants are advanced by single seeds from one generation to the next is referred to as single seed descent method.

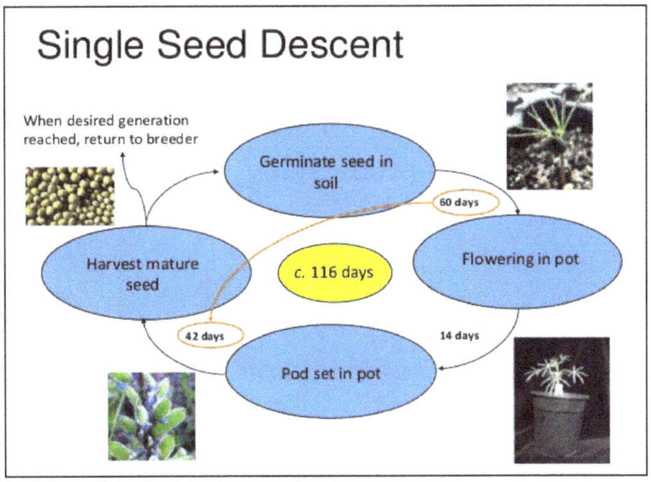

The procedure is as follows:

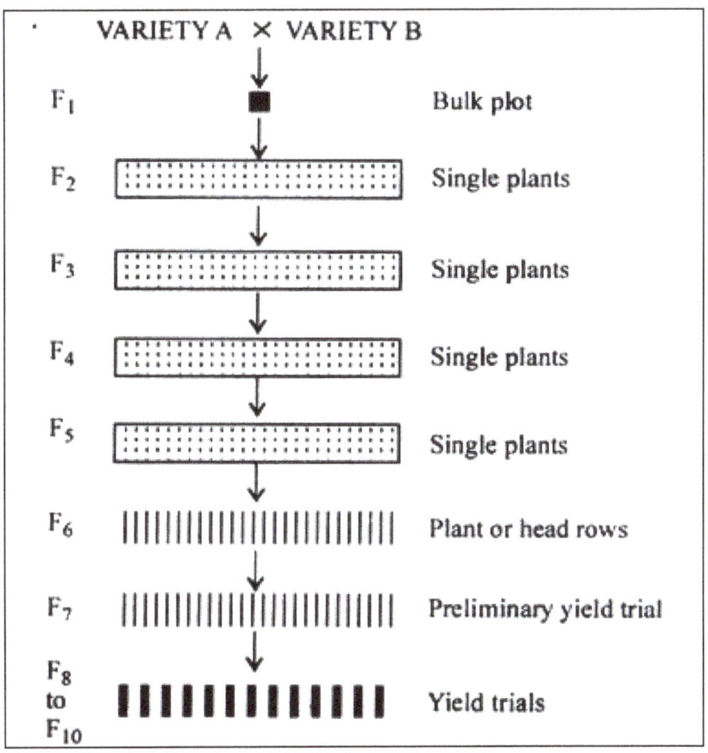

Figure: Single seed-descent method of selection.

Advantages

- It is an easy and rapid way to attain homozygosity.
- Small spaces are required in early generations to grow the selections.
- Natural selection has no effect.

- The duration of the breeding program can be reduced by several years by using single seed descent.
- Every plant originates from a different F2 plant, resulting in greater genetic diversity in each generation.
- It is suited to environments that do not represent those in which the ultimate cultivar will be commercially produced.

Disadvantages

- Natural selection has no effect.
- Plants are selected based on individual phenotype not progeny performance.
- Inability of seed to germinate or plant to set seed may prohibit every F2 plant from being represented in the subsequent population.
- The number of plants in the F2 is equal to the number of plants in the F4. Selecting a single seed per plant runs the risks of losing desirable genes. The assumption is that the single seed represents the genetic base of each F2. This may not be true.

Back Cross Method

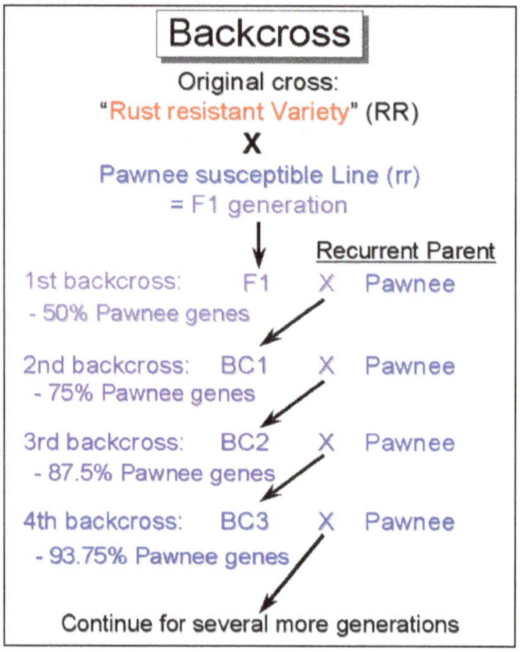

The backcross procedure is most often used to move a transgene from a good tissue culture variety that was used in transformation to an elite experimental line or variety. It turns out that for many crops, once the transgene is in the crop species crossing is more efficient than transformation procedures. Backcrossing is more efficient than transforming the elite line because most transformation protocols are optimized for a specific (often poorly adapted and lower yielding) laboratory line. Many elite lines (which are high yielding) are not amenable for transformation. Hence genetic engineers transform their lab line and breeders backcross the transgene from the lab line into the elite line.

The backcross method is a form of recurrent hybridization (repeated crossing to a single variety) where a superior characteristic may be added to an otherwise desirable variety. In this method the breeder has considerable control of the genetic variation in the segregating population in which selections are to be made. The backcross method has been used extensively for transferring qualitative characters (characters with clear phenotypes that are easy to identify in cross progeny) such as disease resistance. It is effective in both self and cross pollinated crop species. To better understand the applications of backcrossing, the gene for leaf rust resistance in wheat will be used as an example. Figure shows the visible symptoms of leaf rust in susceptible wheat. The left picture leaves are resistant, they have a small amount of rust. The plant on the right is covered with rust, it is susceptible.

Figure: Two examples of wheat, the one on the left is resistant to rust, the one on the right is susceptible to rust.

Figure: illustrates how the backcross procedure can be used to move leaf rust resistance (RR, Rr) from one variety to a susceptible variety (rr).

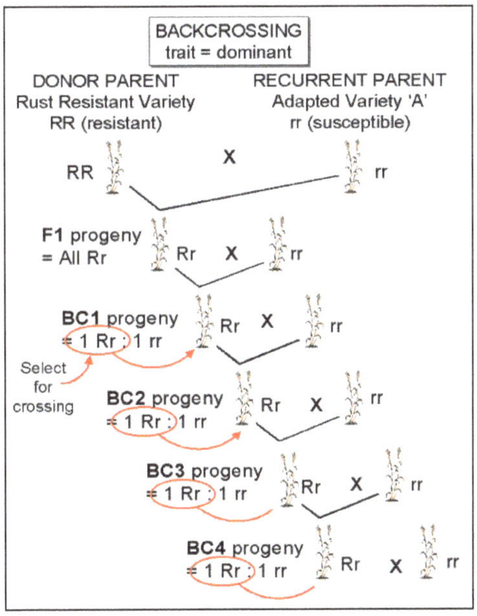

Figure: Backcross breeding with a dominant trait.

The actual procedure for back crossing is almost self-explanatory. In back crossing you have a donor parent (has a gene of interest) and a recurrent parent (an elite line that could be made better by adding the gene of interest). The donor parent is crossed to the recurrent parent. The progeny of this cross is then crossed to the recurrent parent (it is 'crossed back' to the recurrent parent, hence the term back cross). The progeny of this cross is selected for the trait of interest and then crossed back to the recurrent parent. This process is repeated for as many back crosses as are needed to create a line that is the recurrent parent with the gene of interest from the donor parent. The goal of backcrossing is to obtain a line as identical as possible to the recurrent parent with the addition of the gene of interest that has been added through breeding.

In the end, you want to keep only the individuals homozygous for the resistance gene. To obtain them, self Rr plants from BC_4. The resulting offspring will be 1RR : 2Rr : 1rr. Progeny testing would be needed to identify RR from Rr plants. Progeny testing is where the genotype of a parent plant is determined by genotypes of the line's progeny. In the case of an RR plant, the progeny will all be RR (no segregation for the gene/ trait). However in the case of an Rr plant, the progeny will segregate 1/4 RR : 1/2 Rr : 1/4 rr. Therefore, the progeny of RR plants will be uniformly resistant to leaf rust, while the progeny of Rr plants will segregate for resistance and susceptibility.

In contrast, if the genes for rust resistance had been recessive (i.e., ss = resistant) rather than dominant, then the introduced resistant gene is only carried in the heterozygote and would not be detected throughout the backcross program. After each backcross, one would have to self the heterozygote (Ss) in order to produce resistant plants (ss) in the progeny. These resistant plants (ss) are then backcrossed to the recurrent parent (SS).

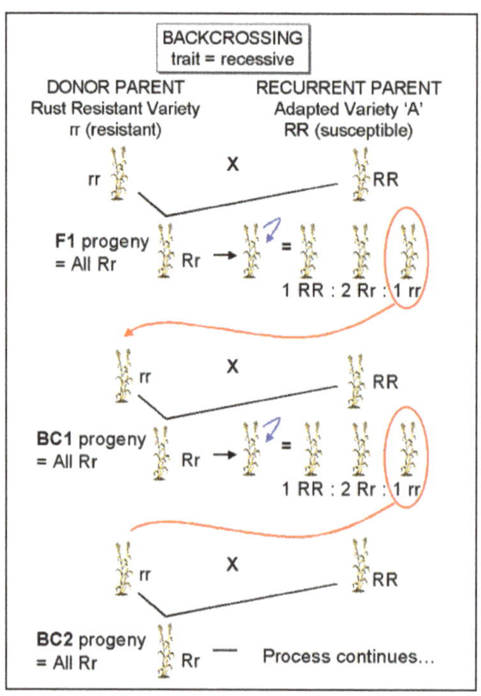

Figure: Backcross breeding with a recessive trait.

When working with recessive traits, such as this example, Allard suggests advancing the 1st backcross to the F_2 generation followed by selection for the desirable character from the donor parent (ss) and the general features of the recurrent parent. The 2nd and 3rd backcrosses are then made in succession after which the inbreeding with selection phase for ss is repeated. This is followed by the 4th and 5th backcrosses in succession. The BC_5F_2 that are resistant (SS) are crossed to recurrent parent (SS) for the BC_6F_1 which is Ss. The BC_7F_1 is selfed to get in the BC_6F_2 : 1/2 SS (susceptible) : 1/2 Ss (susceptible) : 1/2 ss (resistant) backcross with intense selection for both the desired character (ss) and the recurrent parent plant phenotype. You have successfully transfered the gene.

Applications

Generally the backcross method is used for transferring disease resistance character to a good and well adapted variety. Other quantitative characters can also be transferred through back-cross method in both cross- and self-pollinated crop.

1. Inter-varietal Transfer of Simply Inherited Characters

The characters like disease resistance, seed color, plant height, etc., which are controlled by one or two major genes are most suited for transfer through back-cross method from one variety to another of the same species. The successful transfer depends on the minimum linkage between desirable and undesirable trait.

2. Inter-varietal Transfer of Quantitative Characters

Grain characters like seed size and shape, earliness, plant height-all can be transferred from one variety to another with the criteria of high heritability.

3. Interspecific Transfer of Simply Inherited Characters

Mainly the character like disease resistance can be transferred from related species to cultivated species. This specially requires the relatedness between the species where the chromosomes can pair during meiosis. The transfer will be unsuccessful if the genetic environment of recurrent parent is not suitable for functioning of the gene of desirable character from donor parent.

4. Transfer of Cytoplasm

In case of transfer of male sterile character from one parent to another requires backcross method. The variety or species from which the cytoplasm is to be transferred is used as female parent. The recurrent parent should be the male parent. After 6-8 back-crosses, the cytoplasm will be of the donor parent with the genotype of recurrent parent.

5. Transgressive Segregation

Modification of backcross method will produce transgressive segregants. Few (1 to 3) back-crosses with F_1 allow much heterozygosity to appear, or two or more recurrent parents may be used in back- cross programme to accumulate genes from them into the back-cross progeny. Such kind of modification would produce new variety which won't be like the recurrent parent.

6. Production of Isogenic Lines

Backcross method is useful for production of isogenic lines, i.e., the lines of a crop, which are identical in their genotype except for one gene. These isogenic lines are useful for studying the effect of individual genes.

Merits

1. This method does not change the genotype of the popular established variety, only it helps a single desirable character to be transferred in the existing variety.
2. As the recurrent parent is an established variety, so it is not necessary to test the yield performance which ultimately saves five years-time, as well as expenses.
3. In case of short duration plant, 2-3 generations can be raised within a year as the selection is based on inheritance of a particular character, not on performance. So this method drastically reduces the time period required to develop a new variety.
4. Backcross method requires smaller populations than the pedigree method.
5. Defects of an established variety can be removed using this method, may be only by introducing a single character.
6. Interspecific gene transfer only can be done through this method.
7. Transgressive segregation can be obtained in case of quantitative characters by modified backcross method.
8. This method is very much useful for cytoplasmic gene transfer to the recurrent parent, i.e., new variety will bear the cytoplasm of donor parent and genotype of recurrent parent.

Demerits

1. The new variety is not superior in performance than the existing variety except the introduction of single character.

2. During transfer of such kind of single desirable gene, sometimes some undesirable genes may get transferred.

3. For introduction of more than one gene controlled character, multiple crossing programme is necessary which is often difficult, time taking and costly.

4. Once a recurrent parent taken in a backcross method may get replaced by another superior variety of high yielding ability.

Achievements

Backcross method is very much useful for transferring of simply heritable character like disease resistance to well adapted popular local variety.

In case of wheat, 'Kalyan Sona' is the popular established variety to which the leaf rust resistance character has been transferred from diverse sources like Robin, KI, Bluebird, Tobari, etc. using back-cross method.

In case of Bajra, Tift 23A, a male sterile line which was susceptible to downy mildew has made resistant through back-cross method.

For interspecific transfer of genes, back-cross method is widely used for crop improvement. Cultivated sugarcane (Saccharum officincirum) is susceptible to pests and disease, crossed with S. spontanium which is resistant. This brings resistance but with undesirable characters like more fibre, low sugar, thin stem, etc. By back-crossing with noble cane these undesired characters are removed.

In cotton, the hybridization between Gossypium hirsutum and G. arboreum yielded highly sterile F hybrid, few tetraploid seeds were obtained. These plants were then back- crossed with G. hirsutum and two varieties have been selected from the back-cross progeny which are now being widely cultivated in Gujarat.

Except the characters of disease resistance other characters can also be transferred by the back-cross method. Such as in cotton, the ginning quality of fibre has been improved by back-cross method from the local variety. By using BD8 (wilt resistant, high spinning value but low ginning out turns) as recurrent parent and Goghari A26 (high ginning out turn) as non-recurrent parent, new variety Vijay was developed.

Multiple Cross Method

A cross-involving more than one inbred line is referred to as multiple cross. It is also known composite cross and is used to combine monogenetic characters from different sources into a single genotype. In this method, several pure lines are crossed together. The selected pure lines are first combined into crosses as A × B, C × D, E × F, G × H and so on.

The F_1 plants are mated together as (A × B) × (C × D) and (E × F) x (G × H). Finally, the F_1 plants of double crosses are crossed with each other to produce hybrids [(A × B) x (C × D)] × [(E × F) × (G × H)]. Further breeding in these hybrids is carried out according to either pedigree or bulk method.

$$A\times B \quad C\times D \quad E\times F \quad G\times H$$
$$AB\times CD \qquad EF\times GH$$
$$ABCD \quad \times \quad EFGH$$
$$ABCDEFGH$$

Merits

1. In self-pollinated crops this method is used when three or four monogenic characters scattered in three or four different varieties are to be combined into one.
2. These crosses generally have wider adaptation.

Demerits

1. These crosses are generally less productive.
2. This method has limited utility except in high risk areas where severe disease damage occurs regularly from a highly specialized disease pathogen.

Hybrid Varieties

The development of hybrid varieties differs from hybridization in that no attempt is made to produce a pure-breeding population; only the F_1 hybrid plants are sought. The F_1 hybrid of crosses between different genotypes is often much more vigorous than its parents. This hybrid vigour, or heterosis, can be manifested in many ways, including increased rate of growth, greater uniformity, earlier flowering, and increased yield, the last being of greatest importance in agriculture.

By far the greatest development of hybrid varieties has been in corn (maize), primarily because its male flowers (tassels) and female flowers (incipient ears) are separate and easy to handle, thus proving economical for the production of hybrid seed. The production of hand-produced F_1 hybrid seed of other plants, including ornamental flowers, has been economical only because greenhouse growers and home gardeners have been willing to pay high prices for hybrid seed.

Recently, however, a built-in cellular system of pollination control has made hybrid varieties possible in a wide range of plants, including many that are self-pollinating, such as sorghums. This system, called cytoplasmic male sterility, or cytosterility, prevents normal maturation or function of the male sex organs (stamens) and results in defective pollen or none at all. It obviates the need for removing the stamens either by hand or by machine. Cytosterility depends on the interaction between male sterile genes ($R + r$) and factors found in the cytoplasm of the female sex cell. The genes are derived from each parent in the normal Mendelian fashion, but the cytoplasm (and its factors) is provided by the egg only; therefore, the inheritance of cytosterility is determined by the female parent. All plants with fertile cytoplasm produce viable pollen, as do plants with sterile cytoplasm but at least one *R*gene; plants with sterile cytoplasm and two *r* genes are male sterile (produce defective pollen).

The production of F_1 hybrid seed between two strains is accomplished by inter planting a sterile version of one strain (say A) in an isolated field with a fertile version of another strain (B). Since strain A produces no viable pollen, it will be pollinated by strain B, and all seeds produced on strain A plants must therefore be F_1 hybrids between the strains. The F_1 hybrid seeds are then planted to produce the commercial crop. Much of the breeder's work in this process is in developing the pure-breeding sterile and fertile strains to begin the hybrid seed production.

Breeding of Cross-pollinated Plants

The transfer of pollen grains form the anther of one flower to the stigma of another flower belonging to the same species or closely allied species is called cross-pollination or Allogamy. When the pollination takes place between two flowers of the same species, it is called xenogamy, while if it occurs between two closely related species, it is called hybridism.

Characteristics of Cross-pollination

Cross pollination by insect.

Cross-pollination usually occurs in plants having unisexual flower. It may also occur in flowers showing male sterile lines e.g. maize, Solanum. Flowers showing different maturation times for stamens and carpels may also show cross-pollination, e.g. sunflower, Magnolia. It may be in bisexual flowers with differential stamens and carpels, commonly known as heteromorphism e.g., Oxalis. Some flowers, where a barrier is created between stamen and carpel also shows cross-pollination e.g. Iris. The flowers showing different genetic makeup are showing the cross-pollination.

Adaptations for Cross-pollination

a. Dicliny or Unisexuality: In unisexual flowers, self-pollination is impossible, so cross-pollination is observed e.g. gourd.

b. Self-sterility : In certain flowers, the pollen grains are sterile in nature due to male sterility and hence they cannot fertilize the egg of the same flower, so they depend on the pollen grains of another flower coming via cross-pollination, e.g. Solanum; maize.

c. Dichogamy: The androecium and gynoecium of a bisexual flower do not mature at the sametime, so self-pollination can never take place and they are of two types: Protandry (The anther matures earlier than the stigma, e.g. sunflower) and Protogyny (The stigma matures earlier to anther, e.g. Polyalthia).

d. Herkogamy: In this case, self-pollination is impossible, because some of the floral parts act as a physical barrier between anther and stigma and thus cross-pollination is favored. In Iris flower, the anthers are extrorse and they are concealed by overlapping or overarching of the petaloid style, which conceals the anther, making self-pollination impossible.

e. Heteromorphism: The flowers of one single species may vary on the basis of the forms of stamens and carpels and accordingly, the flowers may be dimorphic or trimorphic in nature.

f. Homomorphicself incompatibility: It is a rare type of self incompatibility, in which stamens and pistil do not vary in size, but even than, the pollen grains are rejected by the stigma of the same flower. It is found in Oenothera, Aster.

Methods of Cross-Pollination

There are various types of cross-pollination and the agents are discussed below:

1. Anemophily

When pollination is brought upon by wind, it is called anemophily and the flowers are called anemophilous, e.g. paddy, wheat, maize and grasses.

Adaptations of wind-pollinated flowers: These types of flowers show the following characteristics:

1. The flowers are small and not easily seen.
2. The petals are not colored and they are not scented i.e., devoid of osmophores.
3. They are without nectaries.
4. The flowers are aggregated on a long peduncle above the vegetative parts, which makes the process of wind-pollination easy.
5. The sepals and petals are small and not easily seen and sometimes, undifferentiated to form perianth.
6. The accessory whorls do not cover the sexual reproductive organs.
7. The stamens are provided with long filaments with versatile anthers, which are easily cut off by air current.
8. The pollen grains are small, granular, light weight, dry and shaped in huge quantity.
9. The style is also long and that helps in the protrusion of the stigma from the flower.
10. The stigma is large, feathery and branched, which helps in easy trapping of the pollen grains.

2. Hydrophily

When the cross-pollination in a flower takes place with the help of water, it is called hydrophilous and the phenomenon is termed as hydrophily. The water pollinated plants are of two types:

1. Hypohydrogamous: The pollination taking place in completely submerge condition under water e.g. Ceratophyllum.
2. Epihydrogamous: The pollination taking place along the surface of the wateje.g. Vallisneria and Hydrilla.

Adaptations for Water-pollinated Flowers

The hydrophilous flowers show the following characteristics:

1. The flowers are small, inconspicuous, light in weight, helping in floatation.
2. The flowers are not showy, without colored petals, without any fragrance.
3. The floral parts are covered with waxy substance or cutin, which prevent them from getting damaged by water.

4. The accessory whorls, calyx and corolla are small, so the essential floral whorls or androecium and gynoecium are always exposed in water current.
5. The dehiscence of anther is rapid and so the pollen grains are scattered in wider areas in a short time.
6. The pollen grains are small, light in weight, impervious to water and hence carried by water to long distances.
7. The female flowers usually have a short coiled stalk, that reach the water surface by uncoiling.
8. The stigma is provided with bristles, which can easily trap the pollen grains floating in water.
9. The coiled stalk of female flower may recoil again after pollination.

Method of Pollination

In hypohydrogamous flowers like Ceratophyllum, the flowers never comes above the water surface, the male flowers have superior position, which drop the pollen-grains (impervious to water) on to the stigma of the female flower remaining below. The epihydrogamous flowers are always at the water surface, so pollination is always brought about on the water surface, where pollen grains are carried by the water-current from the male flower, up to the stigma. Some flowers, like that of Vallisneria, the flowers usually remain submerged. The small.male flowers on maturity get detached from the spadix and float on the water surface on a boat-like spathy bract. The coiled stalk of the female flower uncoils and the female flowers come to the water surface, their stigmas come in contact with the anthers of the male flower and pollination takes place. After pollination, the stalk of the female flowers recoil again and the flowers again go down underwater.

3. Zoophily

Cross pollination by bird.

When the pollination is brought about by animals, the pollination is called zoophily and the flowers are called zoophilous e.g. Calotropis. The zoophilous flowers in general shows the following characteristics:

1. The flowers are brightly coloured, showy, sweet scented, i.e., with osmophores.
2. They may contain the nectaries.
3. The pollen grain are edible and palatable.
4. The exines of the pollen grains are rough and sticky, which easily get adhered to the body surface of animals.
5. The stigma surface is usually sticky to trap the pollen grains after coming into its contact.

The cross-pollination brought about by insects is called entomophily and the flowers are called entomophilous e.g. Salvia, sunflower, Calotropis. The pollination affected by birds is called ornithophily and the flowers are called ornithophilous. e.g. BignoniaJSterilitzia. When the pollination is brought about by snails and slugs, it is called malacophily and the flowers are called malacophilous e.g. members of Aroideae. When the process of cross-pollination brought about by bats is called chiropteriphily and the flowers are called chiropteriphilous. e.g. silk cotton, Bauhinia.

The top four breeding methods used for cross-pollinated crops. The methods are:

1. Mass Pedigree Method.
2. Inbreeding.
3. Recurrent Selection.
4. Synthetic Varieties.

Mass Pedigree Method

In this method of breeding, the best individuals with desired characters are selected on the basis of phenotypic performance in a source population. Open pollinated seeds of the selected individual plants are divided into two halves. “Second year replicated progeny row trial is conducted using one set of half seeds from each plant.

On the basis of the progeny performance, the best parental individuals are identified. The remnant half seeds from the superior parental plants are mixed and grown in isolation for random mating during the third year.

This method of breeding is equivalent to ear-to-row selection in context of maize originally proposed by C.G. Hopkins at the Illinois Agricultural Experiment Station in 1896 to improve protein and oil content of maize. This method has been named as mass pedigree method by S.S. Rajan in India. This very method is called line breeding when selection is based on progeny tests and a group of progeny lines is composited.

Inbreeding

The mating of individuals more closely related than individuals mating at random is known as inbreeding. The lines produced by continued inbreeding are known as inbred lines. Self-fertilization is the most intense form of inbreeding.

In plant breeding nearly homozygous lines are produced by continued self-fertilization accompanied by selection for five to six generations. This can be used as the method of breeding only in those crops, which do not show any loss of vigor due to inbreeding, like cucurbits.

The three important uses of inbreeding in cross-pollinated crops are as follows:

1. To attain uniformity in plant characters.
2. To improve yield etc. by individual plant selection as in cucurbits in which there is no inbreeding depression.
3. To develop suitable inbred lines in production of hybrids and synthetics.

Synthetic Variety

The term 'synthetic variety' has come to be used to designate a variety that is maintained from open pollinated seed following its synthesis by hybridization in all combinations among a number of selected genotypes, which have been tested for combining ability.

The components of a synthetic variety could be inbred (usually), clones, mass selected populations or various other materials. The component units are maintained so that the synthetic may be reconstituted at regular intervals.

The inbreeds to be used as component lines are chosen on the basis of combining ability tests. The components inbred are crossed in all possible combinations. This inter-crossed seed is called as Syn 0.

Equal quantity of seed from all crosses is composited and the mixture is allowed open-pollination in isolation and seed is harvested. This becomes Syn 1 generation. In absence of reconstitution of a synthetic at regular intervals, the population becomes an open-pollinated variety.

The testing for combing ability is the decisive criterion for a synthetic variety by which it can be distinguished from a conventional variety of a cross-pollinating species, which originates in a continuous selection of individuals and subsequent progeny tests. The greater variability caused by crossing several components with high general combining ability makes the synthetic varieties more adaptable compared to conventional varieties.

Similar to hybrids, the yield of a synthetic variety generally also decreases after the Syn 2, until an equilibrium is reached which, in partially self-fertile species, depends on selfing rate and inbreeding (minimum depression), but also on the number of components used in the Syn 0.

The performance of Syn 1 can be estimated by the formula:

$$\hat{F}_2 = \overline{F}_1 - \left(\frac{\overline{F}_1 - \overline{P}}{n} \right) 8$$

The yield to be expected, usually increases with the number of components until an optimum is reached. The question regarding the most favorable number of genotypes the Syn 0 should be composed of cannot be clearly answered, because the evidence from research and practice is too divergent.

The yield can be increased by:

1. An increase of the mean performance of the F1 combinations.
2. An increase of the mean performance of the parents (inbred lines, clones).
3. An optimum combination of the components.

Therefore, it is obvious that I_0 plants should already be tested for their combining ability and plants or lines should be used as components, the inbreeding depression of which is not as strong as in the I_5. In order to maintain performance in subsequent generations, mass selections have been found to be sufficient in maize.

The performance of synthetics can be improved by one further breeding cycle. It consists of selection of genotypes from the synthetic variety, their testing by a dialled, and combination of genotypes with the highest combining ability for a new synthesis (= recurrent selection).

Composite Varieties

Concept of composite varieties (in maize) originated in India. Composite varieties are generally derived from the varietal crosses in advanced generation. These are usually developed from open-pollinated varieties or other heterozygous populations or germ plasm which have originally not been subjected to inbreeding or have not been elaborately tested for their combining ability.

Usually, they involve open pollinated varieties, synthetics, double crosses, etc., selected for yield performance, maturity, resistance to diseases and pests. These composites often show a high order heterosis in F_1's when widely diversed populations are crossed. Advanced generations of such heterotic crosses often show stabilized yields. General combining ability and additive gene effects play predominant role in exploitation of these populations.

The details of the steps involved in development of composite variety are as follows:

1. Screening of diverse germ plasm by evaluation at multi-locations/years to identify the sources having adaptability, desirable agronomic attributes and resistance to major diseases and tolerance/resistance to serious insects.
2. Making of all possible crosses among selected superior genotypes or top crossing with a varietal complex of screened base varieties.
3. Conducting multi-location test with the F_1 and F_2 generations of varietal crosses and selection of F_2's showing desirable agronomic features along with least decline in F_2.
4. Evaluation of selected F_2 populations and identification of the best one as practically composites are constituted by compositing seeds of various populations and allowing the mixture to stabilize under open pollination along with some mild selection in isolation. The constituent entries may not be maintained for reconstituting the composite. Composite may serve as a base population for developing inbred lines.

Hybrids

The term hybrid variety is used to designate F_1 populations that are used to commercial planting. The F_1's are obtained by crossing genetically unlike parents. The pioneering work on hybrid maize was done by G.H. Shull (single crosses) and D.F. Jones (double crosses). Other innovative researchers in this area have been E.M. East, H.K. Hayes, F.D. Richey and M.T. Jenkins and others.

Types of Hybrids

(i) Single Cross

A single cross is a hybrid progeny from a cross between two unrelated inbred.

= A x B

(ii) Three-way Cross

A three-way cross is the hybrid progeny from a cross between a single cross and an inbred.

(A x B) x C

(iii) Double Cross

A double cross is the hybrid progeny from a cross between two single crosses.

(A x B) x (C x D)

(iv) Modified Single Cross

A modified single cross is the hybrid progeny from a three- way cross, which utilizes the progeny from two related inbred as the seed parent and an unrelated inbred as the pollen parent.

(A x A') x B

(v) Double Modified Single Cross

A double modified single cross is the hybrid progeny from two single crosses, each developed by crossing two related inbred.

= (A x A') x (B x B')

(vi) Modified Three-way Hybrid

A modified three-way hybrid is the progeny of a single cross as female parent and another single cross between two related inbred.

= (A x B) x (C x C')

(vii) Top cross Hybrid

This is inbred x variety hybrid. Following top cross hybrids may be formed:

a. Inbred line x variety.
b. Inbred line x experimental hybrid.
c. Inbred line x synthetic variety.
d. Inbred line x family.

(viii) Double Top Cross Hybrid

A double top cross hybrid is the progeny of a single cross and a variety. Such hybrids have been produced on commercial scale in India and China.

Methods of Inbred Line Development

Standard Selfing Method

Self-Pollination of individual plants within single plant progenies grown is the most common procedure used to develop inbred lines. This breeding procedure has two important problems.

(a) Vigor of the lines is decreased with inbreeding because of loss of favorable dominant allelles and any heterozygous loci that have over-dominant effects. Many lines are so poor in seed yield, pollen production, etc., that they cannot be used in a programme to produce single cross hybrid seed.

(b) Effective selection within the row for the plants that have desired agronomic traits becomes minimal in generations beyond S_3. Therefore, some breeders use only two or three generations of self-pollination with subsequent reproduction by sib-mating within progenies.

Special Techniques

These techniques involve the doubling of haploids derived from either maternal or paternal gametes. Somaclonal variations from inbred lines offer another opportunity. However, due to low frequency of haploids and doubling of haploids to the diploid state, these methods are still not an important component of most breeding programmes.

Besides, these two methods, there are three other methods where along with developing inbred lines, there are opportunities to improve them simultaneously.

These are as follows:

Pedigree Method

In this method, a pair of elite lines that complement one another are crossed to produce the F_2 generation and pedigree selection is practiced by sampling the F_2 population. F_2 populations of the single crosses are the most frequently used source populations for line development.

Backcrossing

This is a modification of pedigree method. Modifications of backcross method have also been suggested for example, in convergent improvement by Richey, there is parallel improvement of two inbred lines by the reciprocal addition of dominant favourable genes present in one line and lacking in the other line.

In this method, two inbred A and B are crossed. The F_1 is backcrossed with A followed by selection of desirable traits of B and F_1 is also backcrossed with B where selection for desirable traits of A is made. After about three backcrosses and selection, selfing is done to fix the selected genes. This method is useful for improving such characters as vigour, resistance to diseases, pests and lodging.

Gamete Selection

This scheme devised by L.J. Stadler in 1944, is based on the premise that if superior zygotes occur with a frequency of p^2, superior gametes would occur with a frequency of p. The procedure involves crossing an elite line with a random sample of pollen of plants from a source population.

The resulting F_1 plants and the elite line are testcrossed to a common tester and F_1 plants are also selfed. Testcross progenies are evaluated in a replicated trial. The test crosses of F_1 plants that exceed the elite line by tester are presumed to have obtained superior gametes from the source population. Superior gametes are recovered as F_2 self's.

Selection and selfing are continued till desirable homozygosity/uniformity is attained. Though gamete selection is not used as extensively as pedigree and backcross methods, it does have some intrinsic features, and consequently, is included in some breeding programmes.

Testing of Inbred

The general combining ability of the inbred is tested by making all possible crosses [n (n – 1)/2] in a dialed fashion or else top cross test is carried out. Specific combining ability is also estimated.

Time of Testing

One procedure is to test the inbred for hybrid performance/combining ability in about the fifth generation of selfing when the number of selected lines is greatly reduced. This breeding system assumes favorable relationships of plant and other traits of inbred lines (traits as selection criteria during inbreeding) with combining ability for grain yield.

A second system of inbred development is based on an evaluation for hybrid performance in the early generation of self- pollination, e.g., testcrosses of the S_0 plants or S_1 lines. Genotypes that are identified for above average performance in these tests are continued in the selfing and selection nursery.

This procedure has been called as 'early testing' originally proposed by M.T. Jenkins in 1935. The assumption is that the combining ability of a line is determined early in its development and will change relatively little in subsequent generations of inbreeding and selection.

By early testing, the breeder is in a position to discard some lines that are inadequate in hybrid performance and wasteful expenditure on these lines is avoided. However, probably most breeders use a method that is intermediate between these two systems.

In this approach, first hybrid evaluation is of S_2 or S_3 lines.

Combination of Inbred in Hybrids and Prediction of Double Cross Performance:

Single Crosses: n (n – 1) / 2

Three-way crosses: n (n – 1) (n – 2) / 2

Double crosses: n (n – 1) (n – 2) (n – 3) / 8

where, n = Number of inbred

If there are 4 inbred, A, B, C and D, the performance of the double cross (AxB) (CxD) is predicted as follows:

i. Based on mean performance of all possible six single crosses:

   (A x B) + (A x C) + (A x D) + (B x C) + (B x D) + (C x D)/6

ii. Based on mean performance of the four non-parental single crosses:

   (A x C) + (A x D) + (B x C) + (B x D)/4

iii. Based on mean performance of four lines over a series of single crosses:

   (A x E) + (A x F) + (A x G) + (A x H) + (B x E) + (B x F) + (B x G) + (B x H)+.......... +(D x H)/n

iv. Based on mean performance of top-crosses of the four inbreeds:

   (A x variety) + (B x variety) + (C x variety) + D x variety/4

Of the above four methods, method (ii) is found to give more accurate results.

Genetical Basis of Heterosis

The phenomenon of hybrid vigour, expressed particularly in the first generation (F_1) following the crossing of cultivars or inbred lines, has been known for more than a hundred years. The term heterosis, coined by G.H. Shull in 1909 suggests a mechanism based on heterozygosity and therefore, is not fixable in the homozygous state. Two hypotheses have been put forward to explain heterosis.

They are as follows:

(i) Over-dominance Theory

This was proposed by G.H. Shull and E.M. East independently in 1908. According to this hypothesis, hybridity/heterozygosity is superior to either homozygote and this state of heterozygosity has a stimulating effect upon the physiological activities of the organism leading to superiority of Aa over AA or aa.

(ii) Dominance Theory

This hypothesis was proposed by C.B. Davenport in 1908, A.B. Bruce in 1910 and F. Keeble and C. Pellew in 1910. According to this hypothesis, each dominant allele contributes equally to heterosis and the recessive alleles contribute nothing. It is also assumed that the dominance is complete.

If all the dominant alleles are concentrated in one parent, and the counterpart recessive alleles in another parent, the F_1 will be equal to the parent having all the dominant alleles.

For example:

AABBCCDD × aabbccdd
P_1 ↓ P_2
F_1 AaBbCcDd

In the above cross, if each dominant allele, contributes 1 unit and the recessive allele, 0 unit, then the P_1 will have a value of 4 and P_2, a value of 0. F_1 will have a value of 4.

However, there are situations, where F_1 is superior over the better parent.

These cases under dominance theory can be explained assuming that dominance and recessive alleles are distributed in both the parents as given below:

AABBCCdd × aabbccDD
P_1 ↓ P_2
F_1 AaBbCcDd

In this cross, the phenotypic value of P_1 is 3, that of P_2 is 1 and F_1 has a score of 4 which is superior to the better parent. Under this model it should be possible to derive a pure line from the F_1 which should be equal to F_1 in the performance and thus this heterosis will be fixable.

However, it is generally agreed that heterosis is not fixable in the homozygous state. This would be the case if hybrid vigor were due to true over-dominance or due to tight linkage in the repulsion phase at some incompletely dominant loci.

Much evidences suggest that apparent over-dominance is, in fact, due to non-allelic interaction and linkage disequilibrium and that heterosis is mainly a result of the bringing together of unidirectionally dominant alleles distributed between the parental line. Under this, heterozygosity is not an essential prerequisite for high performance, uniformity and stability of performance.

Recurrent Selection

Recurrent selection is a method of breeding designed to concentrate favorable genes scattered among a number of individuals by selecting in each generation among progeny produced by matings inter-se of the selected individuals (or their selfed progeny) of the previous generation.

Based on the ways in which plants with desirable characters are identified, recurrent selection has been divided into four types.

These types are:

1. Simple recurrent selection or recurrent selection for phenotype.
2. Recurrent selection for general combining ability.
3. Recurrent selection for specific combining ability.
4. Reciprocal recurrent selection.

In simple recurrent selection a number of plants are self-pollinated in a source population in first year. At maturity superior plants based on phenotypic performance are selected. In second year, seeds produced by self-fertilization of the selected plants are planted and crossed in all possible combinations and the produce is bulked.

This completes original selection cycle. Since selection is based on the phenotype of the plant, it is useful only for characters with high heritability. In those cases, where it is possible to identify the desired selections before flowering as in case of cauliflower, cabbage, etc., inter-crosses of selections may be made in the first year of each cycle and the second year may be eliminated from each cycle.

Thus, strictly speaking, selfing is not an integral component of simple recurrent selection, rather it is done only to prevent crossing from the inferior pollen grains before the plants reach to selection stage.

In recurrent selection for general combining ability, a three year cycle is involved. In first year a number of plants are self-pollinated and crossed to a broad based heterozygous tester stock to identify the S_0plants with good general combining ability. In second year, the crosses are evaluated to identify those that are superior. Self's of first year are kept in reserve.

In third year, the reserve selfed seeds are grown out, inter-crossed in all combinations, and a composite of inter-crossed seeds is used to establish an improved population for further selection. This procedure developed as a direct outgrowth of studies of early testing first proposed by M.T. Jenkins in 1935.

Recurrent selection for specific combining ability was proposed by F.H. Hull in 1945. This method of selection is same as that of recurrent selection for general combining ability except that the tester selected is a narrow base an inbred line. The recurrent selection for general and specific combining ability is equivalent to half sib progeny test.

Reciprocal recurrent selection proposed by R.E. Comstock, H.F. Robinson and P.H. Harvey in 1949 aims at simultaneous improvement of two heterozygous and heterogenous populations (designated as A and B).

A serves as tester for B and B serves as tester for A. This method is as effective as recurrent selection for gca when additive gene action predominates, and is as effective as recurrent selection for sea when non-additive effects are of major importance.

The steps are as follows:

I. Season:

Selected plants of population A are self-pollinated and crossed to plants of population B. Likewise plants are selected and self-pollinated in B and outcrossed to plants of population A.

II. Season:

Test cross progenies of both the populations are evaluated in replicated trial. Superior progenies are identified on the basis of performance in this trial.

III. Season:

Selfed seed from plants with superior test cross progenies are grown population wise separately and inter-crossed to reconstitute two populations, which will be now called as A' and B'. This completes one cycle and additional cycles may be initiated.

Synthetic Varieties

Synthetic varieties are open pollinated populations developed through random mating of selected genotypes. Synthetic varieties were first suggested by Hayes and Garber and defined by Lonnquist as open pollinated varieties (OPVs) derived from the intercrossing of selfed plants or lines known to possess high general combining ability (GCA), and subsequently maintained by routine mass selection procedures from isolated plantings.

Ultimately the commercial cultivars may be open-pollinated populations (synthetics or composites) and hybrids. The choice of cultivar depends upon resources, stage of breeding program, infrastructure and manpower for seed production and socioeconomic factors. Hybrid cultivars have the advantage of higher yield potential and uniformity. They are preferred over open-pollinated populations subject to higher level of heterosis. However, seed production of hybrids is costlier and a bit tedious and complicated.

Synthetics and the specialized populations derived from them-known as synthetic cultivars (also commonly referred to as synthetic varieties, which are considered completely equivalent to synthetic cultivars here) are common products of plant breeding activities in a wide array of cross-pollinated species. Various definitions have been applied to these populations and some plant breeders have considered them to be equivalent, although this can lead to confusion. Following Lonnquist, a synthetic is an open-pollinated population maintained in isolated plantings that is derived from the random mating of selfed plants or lines or other genotypes (parents) produced from mass selection. As such, a synthetic is simply the bulked seed resulting from one or more cycles of population improvement that involve artificial selection.

Lonnquist defined a synthetic as an open – pollinated population formed by intercrossing of selfed plants or lines and subsequently maintained by mass selection. The term synthetic variety has come to be used to designate a variety that is maintained from open-pollinated seed following its synthesis by hybridization in all possible combinations among a number of selected genotypes which have been subjected to combining ability test. The components of a synthetic variety could be inbreds (usually) or mass selected populations in context of maize. The components are maintained so that the synthetic variety could be reconstituted as and when required.

Developing synthetic varieties through the use of full-sib and half-sib families or clones as parents is a commonly used breeding method in alfalfa. Synthetic varieties have become increasingly favored in alfalfa and other forage crops, mainly because it is cheaper than the development and use of hybrid varieties. Developing synthetic varieties also helps to minimize productivity loss with advancing generations of seed increase.

A synthetic variety is developed through intercrossing of several genotypes of known superior combining ability. Genotypes selected for synthetic variety development are those that are known to give superior hybrid performance when crossed in all combinations. Thus, properly selected male and female parents from diverse origin that can increase the possibility of heterosis when crossed are essential to successfully develop synthetic varieties. Therefore, before selecting clones for developing synthetic varieties, breeders should test the clonal progenies from polycross in yield trials.

In maize first inbreds are developed. The inbreds to be used as component lines are selected on the basis of combining ability for which component inbreds are crossed in all possible combinations. The intercrossed seed is called as So seed. Equal quantity of seed from all crosses is mixed and the mixture is allowed open-pollination in isolation and the seed is harvested. The harvested seed represents S1 generation. In absence of reconstitution of a synthetic variety at regular intervals, the synthetic variety becomes an open-pollinated variety.

The overall performance of the synthetics depend mainly on the number of parental components, their general combining ability, their specific combining ability and on the total amount of heterosis. Synthetics can be used by either farmers for commercial production or breeders as source populations from which to select new lines. To utilize part of the heterosis in faba bean, synthetic cultivars were recommended for spring beans in Europe; the advantages of these cultivars are not only their partial use of heterosis, but also the possible increase in yield stability.

To utilize and increase the involvements of synthetic varieties in crop improvements full information of this issue is very important. But still there is no organized information on the synthetic population, their source, selection methods, developments and their application in agriculture. So this review is objected on providing of detailed materials on the above mentioned issues.

Synthetic Populations

Improved varieties are synthetic cultivars, usually obtained through three or four generations of open pollinated reproduction of polycross seeds of selected parents. This is a great way to create diverse new landraces and open-pollinated lines. Synthetic populations do better than regular open-pollinated types but are not as high yielding as F1 varieties.

Synthetic populations and the composite crosses described below are generally ways to generate diversity from which you can select, either by bulking or recurrent mass selection. Synthetics may also be bred as a goal in itself and the first initial crosses grown together in a mixture that is remade each year (or after several years). To create a synthetic population:

1. Start with two or more variable landraces or heirloom varieties.
2. Make all possible hybrid combinations between all plants. For example, if you started with Brandywine and Green Zebra tomatoes, you would make a hybrid using Brandywine as the mother and Green Zebra as the father (pollen source) and vice versa.
3. Pool all of the seeds of the hybrids together and plant them out.
4. Allow these plants to naturally and randomly homogenize and mix.
5. Over the next three to five generations, select out the best plants.

In obtaining synthetic varieties in mixed species, the reasoning is broader than in allogamous varieties. In the latter, the general and specific combination capacity matter. In mixed species, maintenance of a synthetic variety involves crosses and natural self-pollination. The ideal is thus to select superior inbred lines derived from parents that also have good combination capacity.

Synthetic variety breeding is most effective and intensive method to improve perennial forage crops like alfalfa through polycross. Classical breeding studies require long time to select individual clones for synthetic variety production. Synthetic varieties are populations that are generally created by intermating a set of proven inbred lines.

Synthetics can achieve higher yield levels than older open-pollinated varieties that have received little systematic breeding for yield. They can also be propagated for many generations with little loss of yield. Several studies have shown that synthetics currently produce lower yields that are probably not acceptable to most farmers. The open-pollinated synthetic populations developed for breeding purposes generally produce lower yields, often one-third less than conventional hybrids. Some of these populations do have a high protein content in their grain (9-13%) relative to conventional hybrids (7-9%) so that protein and essential amino acid yields may be similar on a per acre basis. Such open-pollinated corn should have a higher feeding value and price as organic protein from other sources is expensive.

A synthetic variety is developed by intercrossing a number of genotypes known for superior combining ability with high genetic distance. Therefore, synthetic variety is made up of genotypes previously tested for their ability to produce a superior progeny when crossed in all combinations in agreement with Ferreira et al. (1995) who emphasised that heterosis and the combining ability of parents depend directly on the genetic diversity between them and the chance of finding promising combinations is better when more divergent material is used.

Populations can be a result of crosses among a set of homozygous inbred lines (synthetic varieties), an open-pollinated variety, or a mixture of varieties and races (composites). General theories, however, make no distinction about the origin of the population unless it does not fill some of the basic requirements. Synthetic variety constitutes a poly morphical and stable population. Hence synthetic varieties are a high adaptation to environment variations. In other words synthetic varieties provide stable yield in the fluctuating environment.

Sowing seed of synthetic varieties is a common practice in forage species such as alfalfa (Medicago sativa L.) and orchargrass (Dactylis glomerata L.). Such varieties from selection and crossing of lines are phenotypically uniform but different genotypes. These lines to cross freely year after year to produce seeds, heterozygous and heterogeneous populations originate. The use of artificial seed allows multiplication of outstanding genotypes and genetically uniform, since this method does not require that annually crosspollination is carried out to produce plants (McKersie and Brown, 1996). Having the advantage over other method of stable yield and less cost of production of new varieties synthetic population is very unique to be used by breeder in crop improvement.

Sources of Synthetic Populations

Synthetic variety is the same to hybrid once but development of synthetic varieties requires the development of inbred lines, their testing for general combining ability, making their all possible cross combinations, predicting the F2 performance constituting a number of experimental synthetics, testing their yield levels in yield trials over locations, and finally releasing those which excel the standard checks. In India, one of the experimental synthetic populations "Syn 65" has been released for cultivation in the name of cv. "Sangam." In lotni brown sarson (B. campestris), Pusa Kalyani has been developed and released as cultivar from the IARI research station Kanpur (UP) utilizing this breeding approach.

Examining the relationships among progeny tests (correlations, heritability, components of additive and genetic variances) is of great importance in determining which of the tests is the most suitable (effective) for the purposes of breeding and developing synthetic varieties.

Maize (*Zea mays* L.) synthetic populations are low-cost and stable varieties, obtained by cross pollination of a group of inbred lines. They are a viable alternative for situations where the use of hybrid seed and related inputs are too expensive. Although synthetic populations are generally less productive than heterotic hybrids, their main advantage is that the heterosis does not diminish significantly in F_2. Besides the inbred lines, maize synthetics can be obtained from hybrids or local populations. Obtaining maize synthetics from local populations aims at enriching the gene pool with a large number of valuable genes derived from local population characteristic to some agricultural areas.

These positive aspects come to support the approach promoted by CIMMYT in order to obtain highly productive synthetic populations of maize, which is of great importance mainly in places where the use of hybrid seed is too expensive (especially in developing countries). From that year on, synthetic maize populations have acquired a special importance as objectives of research in the field.

Breeders want to improve synthetic populations of maize for using them in obtaining superior inbred lines necessary for hybridization programmers. The value of any maize population depends on its potential per se and on its combining ability in crossings. The per se value of synthetic populations of maize has been studied for many traits: productivity, earliness, resistance to falling, resistance to Sesamia nonagrioides Lef. and Ostrinia nubilalis.

Knowledge about genetic variability in species is important for optimal use of genetic resources in plant breeding programs. The use of molecular markers especially AFLP (Amplified fragment length polymorphism) markers help to select genetic dissimilarity potential parents for production of synthetics. Whatever the type of material will be used in the development of synthetic population its combining ability is the crucial point to be considered.

Development of Synthetic Varieties

The first objective in the development of synthetic varieties is to increase the gene frequency for specific attributes (Hallauer and Eberhart. The basic concept in the development of synthetic varieties is exploitation of heterosis or hybrid vigor, such as varieties are constituted from general combining ability inbreds. However, heterosis is partially utilized by synthetic varieties because some level of inbreeding takes place to open pollination in later generations. Synthetic exploit more of additive gene action whereas hybrid exploit more non additive (over dominance and epistatic) gene action.

A synthetic variety is a variety produced by crossing in all combinations a number of inbred lines (with high GCA that combine well with each other) and a synthetic variety is maintained by open pollination in isolation. In maize, development of synthetics includes:

- Evaluation of lines on the basis of general combining ability.
- These selected lines are intercrossed in all possible combinations.
- Equal amount of seed from these crosses is composited to constitute a synthetic.
- A synthetic variety is developed by intercrossing a number of genotypes known for superior combining ability with high genetic distance. Therefore, synthetic variety is made up of genotypes previously tested for their ability to produce a superior progeny when crossed in all combinations. Though the end goal of development of synthetic varieties is a plant with a general combining ability, following of some stapes and having full knowledge of all this becomes valuable issue. So breeders should ready before starting of the method of improvement.

Selection Methods for Synthetic Variety

After the development of the synthetic population different methods employed to select the best performing individuals. Mass selection and phenotypic selection are the selection methods for synthetic variety development. Molecular markers have been widely used for genetic diversity studies and marker assisted selection for synthetic variety development.

Selecting inbred lines based on their GCA in defined crosses is used to develop synthetics and improve yield when selection is directly on yield (a trait of very low heritability) has a limited effect. This suggests that while the inheritance of combining ability is a quantitative trait governed by many genes, each has a larger individual effect than would genes that contribute directly to yield. Lines with high GCA presumably have a larger proportion of favorable yield genes which differ from other lines measured with respect to their specific favorable gene complex.

Measurements of GCA can be obtained from multiple top crosses made between a line and various elite inbreds, and these crosses are phenotyped for yield in multiple locations and years. Although specific combining ability (SCA) is more important from the standpoint of obtaining maximum yields in hybrid crosses, GCA is highly important in developing high yielding synthetics. Line performance in top crosses has been shown to be relatively constant after the S1 generation and consequently little or nothing would be gained by additional selfing where the production of synthetic varieties is the goal.

Open pollinated varieties developed by modern plant breeding were historically used as commercial cultivars by farmers, and replaced the older landraces that had been selected directly by the farmers. Later, the best OPVs were selected as source populations for further plant improvement and development of synthetics and modern hybrids. In the 1920s, nearly one thousand cultivars available in the United States Corn Belt were selfed in an attempt to develop useful inbred lines. These were often intercrossed to create synthetics. One of the most popular, Iowa Stiff Stalk Synthetic (BSSS), was developed by GF.

Sprague in the early 1930's by intermating 16 inbreds (Hallauer and Miranda, 1988), and is considered an excellent source population for the selection of inbred lines with high combining ability with other elite inbred lines.

THAD reported that the rate of change in inbreeding in synthetic varieties per generation, is always greater than 1/2, for diploid, equilibrium is reached in one step past SYN 1, i. e., in the SYN 2, involved from 2 to 64 parents. In order to evaluate a synthetic variety we need to predict the yield performance of SYN 2.

Practicing prior recurrent selection in the population should increase the occurrence of more homozygous mother plants with good performance in crosses and negligible inbreeding depression. This should be possible from the wide genetic variation detected. Such a fact will favor obtaining superior synthetic varieties. The amount of seeds of a synthetic cultivar is multiplied by successive generations of random mating without selection; these generations are called Syn-2, Syn-3, and so on.

Applications of Synthetic Varieties

Breeding through synthetic variety is most effective and intensive method to improve perennial forage crops like alfalfa through polycross. Classical breeding studies require long time to select individual clones for synthetic variety production. The use of molecular markers especially AFLP (Amplified fragment length polymorphism) markers help to select genetic dissimilarity potential parents for production of synthetics.

The prime synthetics have confirmed to be a precious resource of genetic variability for disease resistance. Synthetic wheat also showed resistance to leaf blotch, glume blotch, crown rot, yellow leaf spot, leaf blight, powdery mildew, karnal bunt. The vast majority of fruit species are propagated by vegetative means because of the presence of selfin compatibility and breeding cycles very long. The use of synthetic seed facilitates its spread. However, the most useful artificial seed would be in the conservation of germplasm of these species. In countries with large varied conditions like Ethiopia synthetic varieties can play a greater role to overcome the influence global warming by increasing the genetic diversity of the crop. In addition to this it can also stabilize the production and productivity as it can resist many diseases and insects.

Synthetic Genomics

Synthetic genomics has been defined as "the engineering of biological components and systems that do not exist in nature and the re-engineering of existing biological elements; it is determined on the intentional design of artificial biological systems, rather than on the understanding of natural biology". Synthetic biology aims to design and model novel biomolecular components, networks and pathways. These are then applied to rewire and reprogram organisms to provide solutions for various challenges.

One of the goals of synthetic genomics is the preparation of viable minimal genomes which will function as platforms for the biochemical production of chemicals with economic relevance. The production of biofuels, pharmaceuticals and the bioremediation of environmental pollution are expected to constitute the first commercial applications of this new technique. Synthetic biology has enabled the construction of a gene that encodes the same amino acid sequence as the plant enzyme but that is optimized for expression in the engineered microorganism of choice. This method has provided massively parallel throughput which has made it possible to identify and track genetic variation among the various strains, providing insights into why some strains are better than others (BIO, 2013). As the world is shifting toward this science, developing countries should relate themselves for better competence in crop improvement and feed their people who are in hunger of food and nutrition.

Breeding of Hybrid Cultivars

A cultivar is a clone or seed strain selected for a particular trait or traits. The value of hybrid cultivars was discovered through experiments with corn in the 1920/30s. Corn is a cross-pollinated crop. But if a corn plant is forced to self-pollinate by covering the silks (=stigmas) from external sources of wind-blown pollen and then pollinating the silks with pollen from the tassels of the same plant seed can be produced. When grown the next generation the progeny do not grow as vigorously as their parent and over many generations of enforced self-pollinating they become even weaker. This is referred to as inbreeding depression. This can be so severe that the weakest lines fail to produce seed but some inbred lines have adequate vigor to be maintained. It was discovered that if inbred lines were hybridized, in certain cases the progeny were more vigorous (larger, higher yielding) than the starting material from which they have been derived. This is known as either hybrid vigor or heterosis. This discovery eventually led to the development of the hybrid corn industry and subsequently other crops.

Whereas other cultivars breed true and their grain can be used as seed for another generation, this is not true of hybrid varieties. Cross-pollination is of course necessary among the plants (in e.g. a hybrid corn field) for grain to be produced. But if that grain is used as seed for next year's crop it will not breed true, the crop will exhibit tremendous variability in all characteristics (height, maturity etc.). In the case of canola hybrids, this would also extend to quality characteristics such as glucosinolate content, erucic acid content and any other oil quality characters of that hybrid and make the seed from that crop unacceptable. Thus farmers must obtain seed of a hybrid (which is produced under controlled conditions by the seed company) each year. Many vegetables are also sold as hybrids such as tomatoes. Saving some seed for planting next year will reveal the hidden variability of a hybrid. As in other crops, new hybrid cultivars (based on new combinations of inbred lines) are being brought to market all the time.

Top Cross Method

A top cross hybrid is defined as a hybrid that results from cross pollinating a single-cross hybrid or an inbred line with an open-pollinated variety. It is considered to be the quickest means of developing hybrids and it produces excess seeds. In addition, the cost of producing top cross hybrids is comparatively cheaper than conventional hybrid seeds. Since the cost of production is cheaper, the price of seeds is also cheaper; hence, small holder resource poor farmers can afford to buy them.

The decline in grain yield due to recycling top cross hybrid seed is half of that of classical hybrids; nonetheless, they balance well with respect to grain yield. The end result of planting "recycled" or second generation seed was insignificant for OPVs (5%), rigorous for hybrids (> 30% yield loss) and intermediate for top cross hybrids (16% yield loss). Given that many farmers in Ghana recycle seeds set aside from their harvest, the use of non-conventional top cross hybrids could be one approach of enhancing yield levels.

In this method widely used in maize, many inbred lines are chosen and back crossed with the original variety there are two types of top crosses viz, single top cross and double top cross.

In single top cross method, two different inbred lines are crossed with each other and the F, crossed with the original variety.

Individuals with desired traits are selected. In double top cross method F hybrids of two different lines are crossed and the product crossed with one of the parents.

Single Cross Method

A hybrid plant results from a cross of two genetically different plants. The two parents of a single-cross hybrid, which is also known as a F1 hybrid, are inbreds. Each seed produced from crossing two inbreds has an array (collection) of alleles from each parent. Those two arrays will be different if the inbreds are genetically different, but each seed contains the same female array and the same male array. Thus, all plants of the same single-cross hybrid are genetically identical. At every locus where the two inbred parents possess different alleles, the single-cross hybrid is heterozygous.

Plants of a single-cross hybrid are more vigorous than the parental inbred plants. In Figures 2a and 2b, the single-cross hybrid plant and ear are shown with the plants and ears of the parental inbreds. Clearly, the hybrid plant is taller and the hybrid ear is larger. The increase in vigor of a hybrid over its two parents is known as hybrid vigor.

Figure: Corn Plants: Inbred B73 (left), Inbred Mo17 (middle), Single cross B73 x Mo17 (right).

Figure: Corn Ears: Inbred B73 (left), Inbred Mo17(right), Single cross B73 x Mo17 (middle).

Breeders often measure the degree of hybrid vigor of a trait with the following formula:

$$\frac{100(Hyb - MP)}{MP}$$

where, *Hyb* = the value of the trait in the hybrid and

MP = the average (mid-parent) value of the trait in the two parents

For example, in figure the height of the single-cross hybrid is 3.0 m (this equals *Hyb*), the average height of the inbreds is 2.0 m (this equals *MP*), and the value of hybrid vigor is 50%. Hybrid vigor calculated in this way is called mid-parent hybrid vigor. Another type is high-parent hybrid vigor. This is the superiority, expressed as a percentage, of the hybrid over the parent with the better or higher value of the trait. Corn breeders will be successful in increasing hybrid performance if the hybrid vigor of a new hybrid compared to an older hybrid is increased and the two sets of parents have equal performance and/or if hybrid vigor is unchanged but the mid-parent value of the parents of the newer hybrid is superior to that of the parents of the older hybrid.

The genetic basis of hybrid vigor is not completely understood. However, experience has shown that a hybrid produced by crossing two inbreds that are closely related usually will exhibit less hybrid vigor than a hybrid produced by crossing inbreds that are more distantly related.

If a single-cross hybrid is allowed to open-pollinate (pollen is despersed freely), each of the plants grown from the resulting seed will be genetically unique. To understand why this is so, first consider a single locus. All plants of a single-cross hybrid are genetically identical, so at a single heterozygous locus any cross- or self-pollination occuring with open-pollination can be represented as:

$$A_1A_2 \times A_1A_2$$

One-half of the egg cells produced by each plant carries the A_1 allele and one-half carry the A_2 allele. The same is true of the pollen cells. The egg and pollen cells then combine at random during pollination.

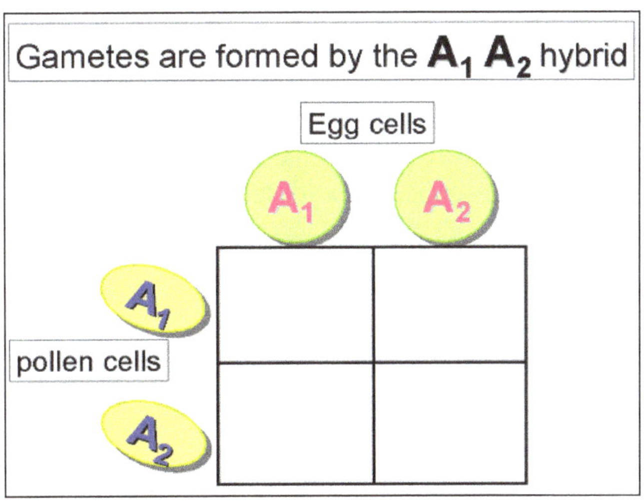

Use the diagram to predict the kinds of offspring produced from this cross.

The A_1A_2 and A_2A_1 genotypes are functionally identical, so three unique genotypes can be produced at a single heterozygous locus when open-pollination occurs.

With two heterozygous loci, *A* and *B*, the situation can be illustrated as follows,

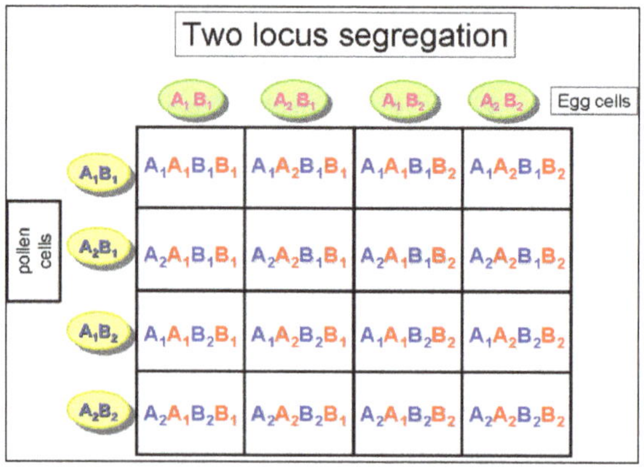

In the two-locus case, nine unique genotypes are produced. This occurrence of multiple genotypes among progeny arising from the self- or cross-pollination of parents that all have the same heterozygous genotype at one or more loci is known as genetic segregation.

This segregation occurs at hundreds or even thousands of gene loci. The number of unique genotypes resulting from genetic segregation at n loci is given by 3^n. Thus, if n=1 (*i.e.*, one locus), then the number of unique genotypes is three, and if n=2 the number of unique genotypes is nine. But, if n=20, the number of unique genotypes balloons to 3,486,784,401 (=3^{20}) Any commercial single-cross hybrid of corn is likely heterozygous at many more than 20 loci. That is why open-pollination of such a single cross results in progeny that are all genetically unique.

The progeny produced from self-pollination of a F1 single-cross hybrid are known as F2 plants. On average, F2 plants will have vigor that is approximately half-way between the single-cross parental plants and the average of the two inbred grandparents; that is, half of the hybrid vigor is lost. This is illustrated in figure. The F2 ears on the bottom row vary in size, but on average are larger than the ears from their inbred grandparents and smaller than the ear from their single-cross parent. That is why farmers have an incentive to purchase new single-cross hybrid seed each year.

Figure: Top row of corn ears: Inbred B73 (left), Single cross B73 x Mo17 (middle), Inbred Mol7 (right) Bottom row of corn ears: From F2 plants derived from B73 x Mo17.

When single-cross hybrid seed is commercially produced, one inbred is the male parent and the other the female parent. Either the female parent must be male-sterile (pollen is not produced or is not functional) or the tassel on each female plant must be removed (this is called detasseling) prior to any pollen production. In either case, all the seed produced on the female parent will be single-cross hybrid seed.

Developing an inbred from a single-cross hybrid requires approximately seven generations of repeated self-pollinations. Each year in the United States, commercial seed companies produce hundreds of new inbreds and test in field trials many thousands of new single-cross hybrids obtained by crossing these inbreds. Compared to existing commercial hybrids, the vast majority of these new hybrids will be poorer or no better in performance. Only the hybrids that have superior performance in these trials are produced in mass quantities and sold as commercial hybrids to farmers.

Figure: A single-cross hybrid production field with female inbred parent detasseled) and male inbred parent .

Considerable time and inputs are required to develop, select, and produce single-cross hybrids. Achieving a high level of cost efficiency of these processes typically requires large-scale operations.

Production of Hybrid Seed

Hybrid is produced by crossing between two genetically dissimilar parents. Pollen from male parent (Pollen parent) will pollinate, fertilize and set seeds in female (seed parent) to produce F1 hybrid seeds. For production of a hybrid CROSSING between two parents is important, the crossing process will results in heterosis. In self pollinated cross it is difficult to cross but in cross pollinated crops it is easier.

In nature to create genetic variability and for its wider adaptation in different environmental conditions, flowering plants has adopted many mechanisms for cross pollination. Cross-pollination results in genetic heterogeneity and show wider adaptations. Flowering plants have evolved a number of devises to encourage cross-pollination. Those mechanisms are:

1. Dicliny: Flowers are unisexual. In monoecious plants male and female flowers are borne on the same plant e.g., cucurbits, maize, castor and coconut. In dioecious plants male flowers are borne on different plants e.g., papaya, cannabis, mulberry.
2. Dichogamy: Time of anther dehiscence and stigma receptivity are different forcing them for cross-pollination. The time gap between the two may vary from one day to many days. In protoandry anthers dehisce earlier than the stigma receptivity e.g; maize, sunflower. In protogyny stigma become recetive earlier than the anther dehisce e.g., Pearl millet mirabilis.
3. Self-incompatibility: Self fertilization in avoided by recognizing the self pollen by the stigma. E.g., Brassica, Petunia, Lilium.
4. Herkogamy: There is spatial separation of the anthers and stigma. Their relative position is such that self fertilization cannot occur. The stigma projects beyond the anthers and therefore pollen cannot land on stigma. E.g., Lucerne stigma is covered with a waxy film. The stigma does not become receptive until this waxy membrane is broken by visit of honeybees resulting in cross-pollination.
5. Male sterility: Absence or atropy or mis or malformed of male sex organ (functional pollen) in normal bisexual flower. Male sterility is of three types: genetic male sterility, cytoplasm sterility and cytoplasmic- genetic male sterility.
6. A combination of two or more of the above mechanisms may occur in some species. This improves the efficiency of the system in promoting cross-pollination.

Among different techniques self-incompatibility and sex expression have significance particularly in vegetable and flower hybrid seed production.

Hand Emasculation and Pollination

Hybrid seeds are produced manually by modifying the plant structure by removal of male organ from female plant before anthesis. This system is possible only when the male and female parts of a single flower or plants are separate. This is being adopted in bisexual perfect flowers where the androecium is removal with case. By removing the anther column / or male part from female line, the sterility of female line is created and is dusted with the pollen of desired male parent.

Self Incompatibility

Self-incompatibility is a mechanism, which avoids self fertilization through recognition of self pollen in or on stigma on the female pistil. But when pollen from other plant carried by wind or insects are accepted and sets seeds.

Self-incompatibility will prevents self pollination (inbreeding) and promotes crosspollination (out breeding) and creates genetic variability. SI is seen in hermaphrodite and homomorphic flowers. Self-incompatibility is a widespread mechanism in flowering plants that prevents inbreeding and promotes outcrossing. The self-incompatibility response is genetically controlled by one or more multi-allelic loci, and relies on a series of complex cellular interactions between the self-incompatible pollen and pistil.

Types of Self-Incompatibility (SI)

Heteromorphic Self-incompatibility

In this system flowers are of different morphology of the reproductive parts. The morphological differences can be seen visibly in flowers this will coincide with crossibility. The characters affecting this type of SI are style length, filament length, pollen size, exine sculpturing. The presence or absence of other SI mechanisms will not affect cross pollination in heteromorphic SI.

- Heteromorphic distyle: e.g., Primula.

The flowers of primula have style at two different heights. There are two types of flowers.

- Thrum flower: short style and long anther.
- Pin flower: has short anthers and long style.

But these two are cross compatible.

- Pin x Pin (ss x ss): Incompatible mating.
- Pin x Thrum (ss x Ss): Compatible mating.
- Thrum x Pin (Ss x ss): Compatible mating.
- Thrum x Thrum (Ss x Ss): Incompatible mating.
    - Heteromorphic Tristyle: e.g., Lythrum.

Three types of flowers:

- Short style and stamens are mid and long.

- Mid style and stamen are short and long.
- Long style and stamen are short and mid.

When homomorphic flowers are crossed it results in incompatible mating and when hetromorphic flowers are crossed it results in compatible mating.

Homomorphic Flowers

Flowers morphological are same, so mating types cannot be recognized by morphological features. This types of self-incompatibility is controlled by alleles. For crossing parental alleles should be different, then only fertilization takes places and seed sets.

Two types of self-incompatibility:

- Sporophytic self-incompatibility.
- Gametophytic self-incompatibility.

Sporophytic Self-incompatibility (SSI)

SSI is less common or rare when compared to GSI. The rejection of pollen is controlled by S loci which is dominant. The dominance relationship are like S1> S2>S3>. This type of self-incompatibility is controlled by diploid genotypes of the sporophyte (pollen). Pollen will not germinate on the stigma of the flower that contains either of the two alleles in the pollen so, rejected.

Gametophytic Self-incompatibility (GSI)

GSSI is more common when compared to SSI. The rejection and acceptance of pollen is controlled by loci. Unlike SSI, incompatibility is controlled by haploid genotype of pollen itself.

S1 pollen can germinate on pistil of S1S2 but due to common S1 allele the pollen tube growth seizes. Similarly in S2 the pollen tube growth seizes. When S3 pollen come in contact with pistil of S1S2 there will be normal growth of pollen tube and fertilization takes place, this is called as partial compatibility. Where as, S3 and S4 pollen can pierce in to the style (S1 S2) and cause fertilization, called as complete compatibility.

In single gene system there are three types of mating systems they are:

- S1S2 X S1S2: 0% compatibility.
- S1S2 X S2S3: 50% compatibility.
- S1S2 X S3S4: 100% compatibility.

Pollen rejection and acceptance of cross pollen is a mechanism of complex interactions.

There are three main interactions:

1. Pollen stigma interaction seen in SSI.
2. Pollen- style interaction seen in GSI.
3. Pollen tube- ovule interaction seen in GSI.

Pollen tube- ovule interaction is very rare and seen in some cases of cocoa, pollen tube reaches ovule and effect fertilization but incompatible combination will degenerate the embryo at early stage development only resulting in no seed set.

Pollen-stigma Interaction

This type of interaction is seen in SSI. It includes dry stigma. The stigma has a hydrated layer of proteins know as pellicle. This pellicle is involved in incompatible reaction. Within a few minutes of pollen reaching stigmatic surface, the pollen releases an exine exudates which is glycoprotein. Due to allele proteins interactions induces immediate callose formation in the papillae, this is in direct contact with pollen. Also this callose is deposited on the protruding pollen tube preventing further germination of the pollen. Here stigma is the site of incompatibility reaction. If once pollen cross this stigmatic barrier, there is no further rejection of pollen.

In case of wet stigma seen in GSI, when pollen sits on the stigma, pollen coat stimulate an unusual form of stigmatic reaction focused beneath the areas of coating. This results in the deposition of fibrogranular electron opaque layer by the extracellular vesicles in the outer layer of the papillae wall. This results in blockage of water passage or loss of osmotic competence by pollen resulting in drying of pollen.

Pollen Tube - style Interaction

This is common in gametophytic self-incompatibility. The pollen can germinate and pollen tube growth will pierce the stigma. But the rate of the pollen tube growth is very slow when compared to compatible pollen. This slow growing pollen tube will stops due to exhaustion of reserve materials and deposition of allele polysaccharide at the tip of the pollen tube which blocks the growth of tube. This deposition is limited to the tip and thus will not affect the other compatible pollen tube growth.

Three-way Hybrid

Two SI lines are crossed to get an heterozygote and again crossed with self-compatible parent which has suppressors locus and give F1 hybrid which is self-compatible.

Triline Hybrid

Hybrid seed production is done by crossing of two inbreed lines. Have to maintenance these inbreed lines which are self-incompatible. For success of self pollination need to eliminate SI. There are many methods to over come SI:

- Bud pollination.
- Washing stigma surface/ pollen grains with organic solvents.
- Mechanical and electric methods.
- CO_2 Treatment.
- High temperature treatment.

Modification of Sex

Hermophrodite flowers has both male and female reproductive organ in a single flower.

- Complete flower; flowers containing all the four whorls viz., sepals, petals, androecium and gyneocium.
- Incomplete flower: flowers missing any of the one whorl.
- Imperfect Flowers- Sexual distinctness: Monoecious (Maize) vs. Dioecious (Hollies, Poplars).
- Perfect Flowers- Gamete Maturation Time (Dicogamy).

Production of this unisexual/imperfect flowers will naturally leads:

- Outcrossing avoids the deleterious effects of inbreeding depression,
- Promotes heterozygosity, genetic variability, and genetic exchange,
- Advantageous to the long-term survival and adaptation of a species.

Monoecious

Flowers are unisexual and are present at different position on the same plant. E.g. cucumber. Terminal flowers are male flower. In the middle of the plant is female favouring crosspollination.

Dioecious: male flowers and female flowers are in different plant. So called as male plant and female plant.

Sex Modification Through Hormones and Chemicals

Sex expression in dioecious and monoecious plants is genetically determined and can be modified to a considerable extent by environmental and introduced factors such as mineral nutrition, photoperiod, temperature, phytohormones. Amongst these, phytohormones have been found to be most effective agents for sex modification and their role in regulation of sex expression in flowering plants has been documented. The morphological differences in various sex types and their specific metabolic characteristics result from the possession of specific patterns of proteins, enzymes and other molecules. Modification of sex expression in cucurbits has been induced both by changing the environmental conditions and by applying treatments with growth regulators. Auxin treatments increase the female sex tendency while gibberellins cause a shift towards maleness.

- Hormones & Chemicals inducing Femaleness: Auxins- NAA, Etherl, Ethephon, Cytokinis- BA, Brassinosteriods etc.
- Hormones & Chemicals inducing Maleness: GA3, AgNO3, ABA Thio porpinic acid, Pthalimide, Paclobutrazol etc.

In cucumber AgNO3 found to be potent inhibitors of ethylene action leading to femaleness. It should be sprayed when first true leaf is fully expanded. Gibberlic acid spray will leads to excessive elongation and weakening of plants and there will be increased number of mall formed male flowers with less pollen. In gynoecious cucumber there will be increased number of male nodes when sprayed with silvernitrate and gibberlic acid, which made possible for multiplication of gynoecious in hybrid seed production.

Environmental Sex Modification

Environment has greater influence on the sex modification. But due to introduction of photosensitive varieties or hybrids in modern era of agriculture it has gained less importance. However in seed production it has its influence on sex expression. In cucumber, high temperature and long day length (> 14 hours) favors male flowers. High temperature will extends the flowering of female flowers. As the temperature increases form 19° C to 23° C the node for first female flowers has also increased from 9.6 to 16.5 number. This clearly indicates that high temperature favors male and delays female flowering.

Male sex expression of several plant species is favored by high temperatures and female sex expression by low temperatures. Male sterile mutant of tomato developing male sterile flowers at a minimum temperature of 30Â°C and normal flowers at lower temperatures. In Brussels sprouts of low temperature effect on the development of the androecium. In onions a slight production of viable pollen by normally male sterile plants above 20 Â°C.

Male Sterility

Hybrid production requires a female plant in which no viable male gametes are borne. Emasculation is done to make a plant devoid of pollen so that it is made female. Another simple way to establish a female line for hybrid seed production is to identify or create a line that is unable to produce viable pollen. This male sterile line is therefore unable to self-pollinate and seed formation is dependent upon pollen from the male line.

In hermaphrodite flowers pollens are non-functional or inactive or sterile while, female gametes functions normally. It is the inability of plant to produce or to release functional pollen as a result of failure of formation or development of functional stamens, microspores or gametes. Male sterility can be either genetic or cytoplasmic or cytoplasmic-genetic. This prevents autogamy and permits crosspollination. Promotes heterozygosity. Sterility is due to nuclear genes or Cytoplasmic gene or both.

In hybrid seed production process female is a male sterile line crossed with male fertility restorer line to get hybrid.

Cytoplasmic Male Sterility

Cytoplasmic male sterility, as the name indicates, is under extra nuclear genetic control mainly mitochondrial genome. They show non-mendelian inheritance and are under the regulation of cytoplasmic factors. In this type, male sterility is inherited maternally. In general there are two types of cytoplasm: N (normal) and the aberrant S (sterile) cytoplasms. These types exhibit reciprocal differences. Cytoplasmic male sterility (CMS) is caused by the extra nuclear genome (mitochondria or chloroplast) and shows maternal inheritance. Manifestation of male sterility in CMS may be either entirely controlled by cytoplasmic factors or by the interaction between cytoplasmic and nuclear factors.

- Stamen (anther and filament) and pollen grains are affected.
- It is divided into:
    - Autoplasmic.

CMS has arisen within a species as a result of spontaneous mutational changes in the cytoplasm, most likely in the mitochondrial genome:

- Alloplasmic.

CMS has arisen from intergeneric, interpecific or occasionally intraspecific crosses and where the male sterility can be interpreted as being due to incompatibility or poor co-operation between nuclear genome of one species and the organellar genome another CMS can be a result of interspecific protoplast fusion. Cytoplasmic male sterility is used in hybrid seed production. In this case, the sterility is transmitted only through the female and all progeny will be sterile. This is not a problem for crops such as onions or carrots where the commodity harvested from the F1 generation is produced during vegetative growth. These CMS lines must be maintained by repeated crossing to a sister line (known as the maintainer line) that is genetically identical except that it possesses normal cytoplasm and is therefore male fertile.

Disadvantages

1. Insufficient or unstable male sterile.
2. Difficulties in restoration system.
3. Difficulties with seed production.

Cytoplasmic-genetic Male Sterility

Male sterility is controlled by an extranuclear genome and often nuclear genes can have the capability to restore fertility. When nuclear restorations of fertility genes are available for CMS system in any crop, it is cytoplasmic-genetic male sterility; the sterility is manifested by the influence of both nuclear (Mendelian inheritance) and cytoplasmic (maternally inherited) genes. There are also restorers of fertility (Rf) genes, which are distinct from genetic male sterility genes. The Rf genes do not have any expression of their own unless the sterile cytoplasm is present. Rf genes are required to restore fertility in S cytoplasm, which causes sterility. Thus N cytoplasm is always fertile and S cytoplasm with genotype Rf- produces fertile; while S cytoplasm with rfrf produces only male steriles. Another feature of these systems is that Rf mutations (i.e., mutations to rf or no fertility restoration) are frequent, so N cytoplasm with Rfrf is best for stable fertility.

Cytoplasmic-genetic male sterility systems are widely exploited in crop plants for hybrid breeding due to the convenience to control the sterility expression by manipulating the gene cytoplasm combinations in any selected genotype. Incorporation of these systems for male sterility evades the need for emasculation in cross-pollinated species, thus encouraging cross breeding producing only hybrid seeds under natural conditions.

In cytoplasmic-genetic male sterility restoration of fertility is done using restorer lines carrying nuclear restorer genes in crops. The male sterile line is maintained by crossing with a maintainer line which has the same genome as that of the MS line but carrying normal fertile cytoplasm.

Genetic Male Sterility

Male sterility is controlled by mutations in nuclear genes in the single recessive genes affect stamen and pollen development, but it can be regulated also by dominant genes. MS alleles are generally recessive. A male sterile line is maintained by crossing with heterozygous male fertile line.

Male sterile plants of monoecious or hermaprodite crops are potentially useful in hybrid program because they eliminate the labor intensive process of flower emasculation Constraint of the use of genetic male sterility.

- The maintenance of the male sterile line. Normally, a GMS line (A-line) is maintained by backcrossing with the heterozygote B-lines (Maintainer lines), but the progeny produced are 50% fertile and 50% male sterile.

Breeding of Clonally Propagated Plants

The definition of a clonally propagated crop is that the material to cultivate and maintain a variety is obtained by asexual reproduction, regardless of how different the plant material used for propagation is within and between species, encompassing tubers, roots, stem cuttings and corms, as well as asexually developed seeds (seeds developed without meiosis). It should be remembered that if crops such as maize (bred as an open-pollinated or hybrid crop) or beans (bred as a cross-fertilized, self-fertilized or hybrid crop) were to be propagated by stem cuttings or asexually developed seeds, they would be clonally propagated crops. In contrast, in breeding clonally propagated crops, the breeding techniques and methods that are usually associated with cross-fertilized and hybrid crops can be very useful. An example is the selection of parents in potato, cassava and sweet potato breeding, which are recombined in open-pollinated polycross nurseries to create new genetic variation. It is almost certain that techniques and methods from breeding cross-fertilized and hybrid crops will become much more important in the future of clone breeding.

It appears to be simple: to break the normal clonal propagation by a crossing step, and thus develop sexual seeds and genetic variation from which to select new clones. All propagation steps from the first to the last selection step are again 'normal' asexual reproduction. Hence, the finally selected clone is genetically identical with the original seed plant from which the selected clone is derived. In other words, each seed plant is a potential variety. Roots and tubers, fruit and tree plant species have been used by human since long before the dawn of agriculture. They have been domesticated by IPB and several made a substantial yield progress by FPB in some regions of the world. However, in other regions of the world there is not much yield progress, and in these regions there appears to be a clear need of PPB for progress.

An example of the needs and requirements of clonally propagated varieties can be found in potato (Solanum spp.). There are about 200 wild potato species. They usually contain glycoalkaloids, which give tubers a bitter taste and which are toxic when consumed in large quantities. It is nearly certain that 100 to 130 centuries ago indigenous knowledge in the Andes and along the Pacific coast of South America was those sites where it was possible to collect wild potato tubers where species and mutants were growing that had low alkaloid content. Although these tubers were very small, the man, or more probably a woman, made life much easier by growing and maintaining desirable types by cloning close to their homes. This happened more than 8 000 years ago, and most likely independently at several places. Those types were preferred that were easier to maintain, easier to harvest (shorter stolons) and had larger tubers compared to other types. The result was the domestication of pitiquiña (Solanum stenotomum), which was most probably selected from S. leptophyes or S. canasense. From the view-point of the knowledge of the twenty-first century it is not surprising that suddenly potato plants with larger leaves and larger tubers were found. Potato spontaneously changes its polyploidy level by unreduced gametes and recombination. Polyploid potatoes are more vigorous than their diploid ancestors. The result was the domesticated of polyploid andigena (S. tuberosum subsp. andigena). Andigena is the ancestor of the commercial potato in long-day temperate climates—the so-called Irish potato (S. tuberosum subsp. tuberosum). This IPB of potato and introductions of FV of potato into the Northern Hemisphere changed the world both socio-economically and politically.

References

- Top-3-methods-used-in-self-pollinated-crops-breeding-methods-60846: biologydiscussion.com, Retrieved 20 May 2020
- Procedure-for-pureline-selection, pure-line-selection, plant-breeding: theagricos.com, Retrieved 25 April 2020
- Bulk-method-merits-applications, plant-breeding: theagricos.com, Retrieved 23 April 2020
- Top-3-Breeding-Methods-Used-for-Cross-Pollinated-Crops-367364157: scribd.com, Retrieved 08 July 2020
- Breeding-self-pollinated-species, plant-breeding: britannica.com, Retrieved 31 March 2020
- Hybridization-method-of-crop-improvement-17701: biologydiscussion.com, Retrieved 17 March 2020

Chapter 5

Different Types of Genetically Modified Plants

Genetic modification in plants has come a long way and has resulted in many genetically modified crops being produced and cultivated commercially. These crops have better traits and qualities which make them much more viable for cultivation, consumption and commercial purposes. The topic of genetically modified plants is covered in detail in this chapter.

Genetically Modified Crops

Genetically modified crops (GMCs, GM crops, or biotech crops) are plants used in agriculture, the DNA of which has been modified using genetic engineering techniques. In most cases, the aim is to introduce a new trait to the plant which does not occur naturally in the species. Examples in food crops include resistance to certain pests, diseases, or environmental conditions, reduction of spoilage, or resistance to chemical treatments (e.g. resistance to a herbicide), or improving the nutrient profile of the crop. Examples in non-food crops include production of pharmaceutical agents, biofuels, and other industrially useful goods, as well as for bioremediation.

Farmers have widely adopted GM technology. Between 1996 and 2013, the total surface area of land cultivated with GM crops increased by a factor of 100, from 17,000 km^2 (4.2 million acres) to 1,750,000 km^2 (432 million acres). 10% of the world's arable land was planted with GM crops in 2010. In the US, by 2014, 94% of the planted area of soybeans, 96% of cotton and 93% of corn were genetically modified varieties. Use of GM crops expanded rapidly in developing countries, with about 18 million farmers growing 54% of worldwide GM crops by 2013. A 2014 meta-analysis concluded that GM technology adoption had reduced chemical pesticide use by 37%, increased crop yields by 22%, and increased farmer profits by 68%. This reduction in pesticide use has been ecologically beneficial, but benefits may be reduced by overuse. Yield gains and pesticide reductions are larger for insect-resistant crops than for herbicide-tolerant crops. Yield and profit gains are higher in developing countries than in developed countries.

There is a scientific consensus that currently available food derived from GM crops poses no greater risk to human health than conventional food, but that each GM food needs to be tested on a case-by-case basis before introduction. Nonetheless, members of the public are much less likely than scientists to perceive GM foods as safe. The legal and regulatory status of GM foods varies by country, with some nations banning or restricting them, and others permitting them with widely differing degrees of regulation.

However, opponents have objected to GM crops on several grounds, including environmental concerns, whether food produced from GM crops is safe, whether GM crops are needed to address the world's food needs, and concerns raised by the fact these organisms are subject to intellectual property law.

Gene Transfer in Nature and Traditional Agriculture

DNA transfers naturally between organisms. Several natural mechanisms allow gene flow across species. These occur in nature on a large scale – for example, it is one mechanism for the development of antibiotic resistance in bacteria. This is facilitated by transposons, retrotransposons, proviruses and other mobile genetic elements that naturally translocate DNA to new loci in a genome. Movement occurs over an evolutionary time scale.

The introduction of foreign germplasm into crops has been achieved by traditional crop breeders by overcoming species barriers. A hybrid cereal grain was created in 1875, by crossing wheat and rye. Since then important traits including dwarfing genes and rust resistance have been introduced. Plant tissue culture and deliberate mutations have enabled humans to alter the makeup of plant genomes.

History

The first genetically modified crop plant was produced in 1982, an antibiotic-resistant tobacco plant. The first field trials occurred in France and the USA in 1986, when tobacco plants were engineered for herbicide resistance. In 1987, Plant Genetic Systems (Ghent, Belgium), founded by Marc Van Montagu and Jeff Schell, was the first company to genetically engineer insect-resistant (tobacco) plants by incorporating genes that produced insecticidal proteins from *Bacillus thuringiensis* (Bt).

The People's Republic of China was the first country to allow commercialized transgenic plants, introducing a virus-resistant tobacco in 1992, which was withdrawn in 1997. The first genetically modified crop approved for sale in the U.S., in 1994, was the *FlavrSavr* tomato. It had a longer shelf life, because it took longer to soften after ripening. In 1994, the European Union approved tobacco engineered to be resistant to the herbicide bromoxynil, making it the first commercially genetically engineered crop marketed in Europe.

In 1995, Bt Potato was approved by the US Environmental Protection Agency, making it the country's first pesticide producing crop. In 1995 canola with modified oil composition (Calgene), Bt maize (Ciba-Geigy), bromoxynil-resistant cotton (Calgene), Bt cotton (Monsanto), glyphosate-resistant soybeans (Monsanto), virus-resistant squash (Asgrow), and additional delayed ripening tomatoes (DNAP, Zeneca/Peto, and Monsanto) were approved. As of mid-1996, a total of 35 approvals had been granted to commercially grow 8 transgenic crops and one flower crop (carnation), with 8 different traits in 6 countries plus the EU. In 2000, Vitamin A-enriched golden rice was developed, though as of 2016 it was not yet in commercial production. In 2013 the leaders of the three research teams that first applied genetic engineering to crops, Robert Fraley, Marc Van Montagu and Mary-Dell Chilton were awarded the World Food Prize for improving the "quality, quantity or availability" of food in the world.

Methods

Genetically engineered crops have genes added or removed using genetic engineering techniques, originally including gene guns, electroporation, microinjection and agrobacterium. More recently, CRISPR and TALEN offered much more precise and convenient editing techniques.

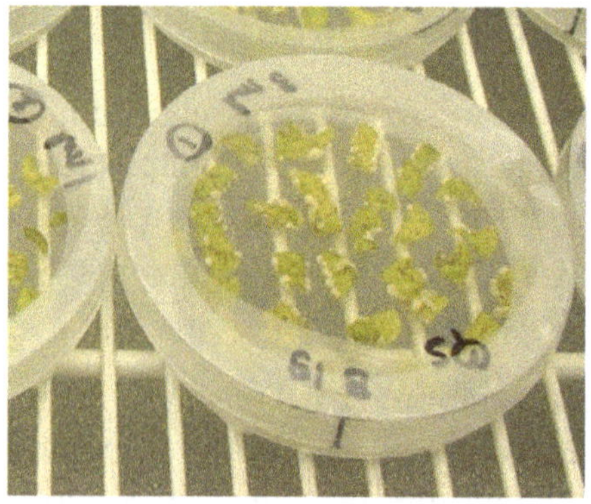

Plants (*Solanum chacoense*) being transformed using agrobacterium.

Gene guns (a.k.a. biolistic) "shoot" (direct high energy particles or radiations against) target genes into plant cells. It is the most common method. DNA is bound to tiny particles of gold or tungsten which are subsequently shot into plant tissue or single plant cells under high pressure. The accelerated particles penetrate both the cell wall and membranes. The DNA separates from the metal and is integrated into plant DNA inside the nucleus. This method has been applied successfully for many cultivated crops, especially monocots like wheat or maize, for which transformation using *Agrobacterium tumefaciens* has been less successful. The major disadvantage of this procedure is that serious damage can be done to the cellular tissue.

Agrobacterium tumefaciens-mediated transformation is another common technique. Agrobacteria are natural plant parasites, and their natural ability to transfer genes provides another engineering method. To create a suitable environment for themselves, these Agrobacteria insert their genes into plant hosts, resulting in a proliferation of modified plant cells near the soil level (crown gall). The genetic information for tumour growth is encoded on a mobile, circular DNA fragment (plasmid). When Agrobacterium infects a plant, it transfers this T-DNA to a random site in the plant genome. When used in genetic engineering the bacterial T-DNA is removed from the bacterial plasmid and replaced with the desired foreign gene. The bacterium is a vector, enabling transportation of foreign genes into plants. This method works especially well for dicotyledonous plants like potatoes, tomatoes, and tobacco. Agrobacteria infection is less successful in crops like wheat and maize.

Electroporation is used when the plant tissue does not contain cell walls. In this technique, "DNA enters the plant cells through miniature pores which are temporarily caused by electric pulses."

Microinjection directly injects the gene into the DNA.

Plant scientists, backed by results of modern comprehensive profiling of crop composition, point out that crops modified using GM techniques are less likely to have unintended changes than are conventionally bred crops.

In research tobacco and *Arabidopsis thaliana* are the most frequently modified plants, due to well-developed transformation methods, easy propagation and well studied genomes. They serve as model organisms for other plant species.

Introducing new genes into plants requires a promoter specific to the area where the gene is to be expressed. For instance, to express a gene only in rice grains and not in leaves, an endosperm-specific promoter is used. The codons of the gene must be optimized for the organism due to codon usage bias. Transgenic gene products should be able to be denatured by heat so that they are destroyed during cooking.

Types of Modifications

Transgenic maize containing a gene from the bacteria *Bacillus thuringiensis.*

Transgenic

Transgenic plants have genes inserted into them that are derived from another species. The inserted genes can come from species within the same kingdom (plant to plant) or between kingdoms (for example, bacteria to plant). In many cases the inserted DNA has to be modified slightly in order to correctly and efficiently express in the host organism. Transgenic plants are used to express proteins like the cry toxins from *B. thuringiensis*, herbicide resistant genes, antibodies and antigens for vaccinations A study led by the European Food Safety Authority (EFSA) found also viral genes in transgenic plants.

Transgenic carrots have been used to produce the drug Taliglucerase alfa which is used to treat Gaucher's disease. In the laboratory, transgenic plants have been modified to increase photosynthesis (currently about 2% at most plants versus the theoretic potential of 9–10%). This is possible by changing the rubisco enzyme (i.e. changing C3 plants into C4 plants), by placing the rubisco in a carboxysome, by adding CO_2 pumps in the cell wall, by changing the leaf form/size. Plants have been engineered to exhibit bioluminescence that may become a sustainable alternative to electric lighting.

Cisgenic

Cisgenic plants are made using genes found within the same species or a closely related one, where conventional plant breeding can occur. Some breeders and scientists argue that cisgenic modification is useful for plants that are difficult to crossbreed by conventional means (such as potatoes), and that plants in the cisgenic category should not require the same regulatory scrutiny as transgenics.

Subgenic

Genetically modified plants can also be developed using gene knockdown to alter the genetic makeup of a plant without incorporating genes from other plants. In 2014, Chinese researcher Gao Caixia filed patents on the creation of a strain of wheat that is resistant to powdery mildew. The strain lacks genes that encode proteins that repress defenses against the mildew. The researchers deleted all three copies of the genes from wheat's hexaploid genome. Gao used the TALENs and CRISPR gene editing tools without adding or changing any other genes. No field trials were immediately planned. The CRISPR technique has also been used to modify white button mushrooms (*Agaricus bisporus*).

Economics

GM food's economic value to farmers is one of its major benefits, including in developing nations. A 2010 study found that Bt corn provided economic benefits of $6.9 billion over the previous 14 years in five Midwestern states. The majority ($4.3 billion) accrued to farmers producing non-Bt corn. This was attributed to European corn borer populations reduced by exposure to Bt corn, leaving fewer to attack conventional corn nearby. Agriculture economists calculated that "world surplus [increased by] $240.3 million for 1996. Of this total, the largest share (59%) went to U.S. farmers. Seed company Monsanto received the next largest share (21%), followed by US consumers (9%), the rest of the world (6%), and the germplasm supplier, Delta & Pine Land Company of Mississippi (5%)."

According to the International Service for the Acquisition of Agri-biotech Applications (ISAAA), in 2014 approximately 18 million farmers grew biotech crops in 28 countries; about 94% of the farmers were resource-poor in developing countries. 53% of the global biotech crop area of 181.5 million hectares was grown in 20 developing countries. PG Economics comprehensive 2012 study concluded that GM crops increased farm incomes worldwide by $14 billion in 2010, with over half this total going to farmers in developing countries.

Critics challenged the claimed benefits to farmers over the prevalence of biased observers and by the absence of randomized controlled trials. The main Bt crop grown by small farmers in developing countries is cotton. A 2006 review of Bt cotton findings by agricultural economists concluded, "the overall balance sheet, though promising, is mixed. Economic returns are highly variable over years, farm type, and geographical location".

In 2013 the European Academies Science Advisory Council (EASAC) asked the EU to allow the development of agricultural GM technologies to enable more sustainable agriculture, by employing fewer land, water and nutrient resources. EASAC also criticizes the EU's "timeconsuming and expensive regulatory framework" and said that the EU had fallen behind in the adoption of GM technologies.

Participants in agriculture business markets include seed companies, agrochemical companies, distributors, farmers, grain elevators and universities that develop new crops/traits and whose agricultural extensions advise farmers on best practices. According to a 2012 review based on data from the late 1990s and early 2000s, much of the GM crop grown each year is used for livestock feed and increased demand for meat leads to increased demand for GM feedcrops. Feed grain usage as a percentage of total crop production is 70% for corn and more than 90% of oil seed meals such as soybeans. About 65 million metric tons of GM corn grains and about 70 million metric tons of soybean meals derived from GM soybean become feed.

In 2014 the global value of biotech seed was US$15.7 billion; US$11.3 billion (72%) was in industrial countries and US$4.4 billion (28%) was in the developing countries. In 2009, Monsanto had $7.3 billion in sales of seeds and from licensing its technology; DuPont, through its Pioneer subsidiary, was the next biggest company in that market. As of 2009, the overall Roundup line of products including the GM seeds represented about 50% of Monsanto's business.

Some patents on GM traits have expired, allowing the legal development of generic strains that include these traits. For example, generic glyphosate-tolerant GM soybean is now available. Another impact is that traits developed by one vendor can be added to another vendor's proprietary strains, potentially increasing product choice and competition. The patent on the first type of *Roundup Ready* crop that Monsanto produced (soybeans) expired in 2014 and the first harvest of off-patent soybeans occurs in the spring of 2015. Monsanto has broadly licensed the patent to other seed companies that include the glyphosate resistance trait in their seed products. About 150 companies have licensed the technology, including Syngenta and DuPont Pioneer.

Yield

In 2014, the largest review yet concluded that GM crops' effects on farming were positive. The meta-analysis considered all published English-language examinations of the agronomic and economic impacts between 1995 and March 2014 for three major GM crops: soybean, maize, and cotton. The study found that herbicide-tolerant crops have lower production costs, while for insect-resistant crops the reduced pesticide use was offset by higher seed prices, leaving overall production costs about the same.

Yields increased 9% for herbicide tolerance and 25% for insect resistant varieties. Farmers who adopted GM crops made 69% higher profits than those who did not. The review found that GM crops help farmers in developing countries, increasing yields by 14 percentage points.

The researchers considered some studies that were not peer-reviewed, and a few that did not report sample sizes. They attempted to correct for publication bias, by considering sources beyond academic journals. The large data set allowed the study to control for potentially confounding variables such as fertiliser use. Separately, they concluded that the funding source did not influence study results.

Traits

GM crops grown today, or under development, have been modified with various traits. These traits include improved shelf life, disease resistance, stress resistance, herbicide resistance, pest resistance, production of useful goods such as biofuel or drugs, and ability to absorb toxins and for use in bioremediation of pollution.

Recently, research and development has been targeted to enhancement of crops that are locally important in developing countries, such as insect-resistant cowpea for Africa and insect-resistant brinjal (eggplant).

Lifetime

The first genetically modified crop approved for sale in the U.S. was the *FlavrSavr* tomato, which had a longer shelf life. It is no longer on the market.

In November 2014, the USDA approved a GM potato that prevents bruising.

In February 2015 Arctic Apples were approved by the USDA, becoming the first genetically modified apple approved for US sale. Gene silencing was used to reduce the expression of polyphenol oxidase (PPO), thus preventing enzymatic browning of the fruit after it has been sliced open. The trait was added to Granny Smith and Golden Delicious varieties. The trait includes a bacterial antibiotic resistance gene that provides resistance to the antibiotic kanamycin. The genetic engineering involved cultivation in the presence of kanamycin, which allowed only resistant cultivars to survive. Humans consuming apples do not acquire kanamycin resistance, per arcticapple.com. The FDA approved the apples in March 2015.

Nutrition

Edible Oils

Some GM soybeans offer improved oil profiles for processing or healthier eating. Camelina sativa has been modified to produce plants that accumulate high levels of oils similar to fish oils.

Vitamin Enrichment

Golden rice, developed by the International Rice Research Institute (IRRI), provides greater amounts of Vitamin A targeted at reducing Vitamin A deficiency. As of January 2016, golden rice has not yet been grown commercially in any country.

Researchers vitamin-enriched corn derived from South African white corn variety M37W, producing a 169-fold increase in Vitamin A, 6-fold increase in Vitamin C and doubled concentrations of folate. Modified Cavendish bananas express 10-fold the amount of Vitamin A as unmodified varieties.

Toxin Reduction

A genetically modified cassava under development offers lower cyanogen glucosides and enhanced protein and other nutrients (called BioCassava).

In November 2014, the USDA approved a potato, developed by J.R. Simplot Company, that prevents bruising and produces less acrylamide when fried. The modifications prevent natural, harmful proteins from being made via RNA interference. They do not employ genes from non-potato species. The trait was added to the Russet Burbank, Ranger Russet and Atlantic varieties.

Stress Resistance

Plants engineered to tolerate non-biological stressors such as drought, frost, high soil salinity, and nitrogen starvation were in development. In 2011, Monsanto's DroughtGard maize became the first drought-resistant GM crop to receive US marketing approval.

Herbicides

Glyphosate

As of 1999 the most prevalent GM trait was glyphosate-resistance. Glyphosate, (the active ingredient in Roundup and other herbicide products) kills plants by interfering with the shikimate pathway in plants, which is essential for the synthesis of the aromatic amino acids phenylalanine, tyrosine and tryptophan. The shikimate pathway is not present in animals, which instead obtain aromatic amino acids from their diet. More specifically, glyphosate inhibits the enzyme 5-enolpyruvylshikimate-3-phosphate synthase (EPSPS).

This trait was developed because the herbicides used on grain and grass crops at the time were highly toxic and not effective against narrow-leaved weeds. Thus, developing crops that could withstand spraying with glyphosate would both reduce environmental and health risks, and give an agricultural edge to the farmer.

Some micro-organisms have a version of EPSPS that is resistant to glyphosate inhibition. One of these was isolated from an *Agrobacterium* strain CP4 (CP4 EPSPS) that was resistant to glyphosate. The CP4 EPSPS gene was engineered for plant expression by fusing the 5' end of the gene to a chloroplast transit peptide derived from the petunia EPSPS. This transit peptide was used because it had shown previously an ability to deliver bacterial EPSPS to the chloroplasts of other plants. This CP4 EPSPS gene was cloned and transfected into soybeans.

The plasmid used to move the gene into soybeans was PV-GMGTO4. It contained three bacterial genes, two CP4 EPSPS genes, and a gene encoding beta-glucuronidase (GUS) from *Escherichia coli* as a marker. The DNA was injected into the soybeans using the particle acceleration method. Soybean cultivar A5403 was used for the transformation.

Bromoxynil

Tobacco plants have been engineered to be resistant to the herbicide bromoxynil.

Glufosinate

Crops have been commercialized that are resistant to the herbicide glufosinate, as well. Crops engineered for resistance to multiple herbicides to allow farmers to use a mixed group of two, three, or four different chemicals are under development to combat growing herbicide resistance.

2,4-D

In October 2014 the US EPA registered Dow's Enlist Duo maize, which is genetically modified to be resistant to both glyphosate and 2,4-D, in six states. Inserting a bacterial aryloxyalkanoate dioxygenase gene, *aad1* makes the corn resistant to 2,4-D. The USDA had approved maize and soybeans with the mutation in September 2014.

Dicamba

Monsanto has requested approval for a stacked strain that is tolerant of both glyphosate and dicamba.

Pest Resistance

Insects

Tobacco, corn, rice and many other crops have been engineered to express genes encoding for insecticidal proteins from Bacillus thuringiensis (Bt). Papaya, potatoes, and squash have been engineered to resist viral pathogens such as cucumber mosaic virus which, despite its name, infects a wide variety of plants. The introduction of Bt crops during the period between 1996 and 2005 has been estimated to have reduced the total volume of insecticide active ingredient use in the United States by over 100 thousand tons. This represents a 19.4% reduction in insecticide use.

In the late 1990s, a genetically modified potato that was resistant to the Colorado potato beetle was withdrawn because major buyers rejected it, fearing consumer opposition.

Viruses

Virus resistant papaya were developed In response to a papaya ringspot virus (PRV) outbreak in Hawaii in the late 1990s. . They incorporate PRV DNA. By 2010, 80% of Hawaiian papaya plants were genetically modified.

Potatoes were engineered for resistance to potato leaf roll virus and Potato virus Y in 1998. Poor sales led to their market withdrawal after three years.

Yellow squash that were resistant to at first two, then three viruses were developed, beginning in the 1990s. The viruses are watermelon, cucumber and zucchini/courgette yellow mosaic. Squash was the second GM crop to be approved by US regulators. The trait was later added to zucchini.

Many strains of corn have been developed in recent years to combat the spread of Maize dwarf mosaic virus, a costly virus that causes stunted growth which is carried in Johnson grass and spread by aphid insect vectors. These strands are commercially available although the resistance is not standard among GM corn variants.

By-products

Drugs

In 2012, the FDA approved the first plant-produced pharmaceutical, a treatment for Gaucher's Disease. Tobacco plants have been modified to produce therapeutic antibodies.

Biofuel

Algae is under development for use in biofuels. Researchers in Singapore were working on GM jatropha for biofuel production. Syngenta has USDA approval to market a maize trademarked Enogen that has been genetically modified to convert its starch to sugar for ethanol. In 2013, the Flemish Institute for Biotechnology was investigating poplar trees genetically engineered to contain less lignin to ease conversion into ethanol. Lignin is the critical limiting factor when using wood to make bio-ethanol because lignin limits the accessibility of cellulose microfibrils to depolymerization by enzymes.

Materials

Companies and labs are working on plants that can be used to make bioplastics. Potatoes that produce industrially useful starches have been developed as well. Oilseed can be modified to produce fatty acids for detergents, substitute fuels and petrochemicals.

Bioremediation

Scientists at the University of York developed a weed (*Arabidopsis thaliana*) that contains genes from bacteria that can clean TNT and RDX-explosive soil contaminants. 16 million hectares in the

USA (1.5% of the total surface) are estimated to be contaminated with TNT and RDX. However *A. thaliana* was not tough enough for use on military test grounds.

Genetically modified plants have been used for bioremediation of contaminated soils. Mercury, selenium and organic pollutants such as polychlorinated biphenyls (PCBs).

Marine environments are especially vulnerable since pollution such as oil spills are not containable. In addition to anthropogenic pollution, millions of tons of petroleum annually enter the marine environment from natural seepages. Despite its toxicity, a considerable fraction of petroleum oil entering marine systems is eliminated by the hydrocarbon-degrading activities of microbial communities. Particularly successful is a recently discovered group of specialists, the so-called hydrocarbonoclastic bacteria (HCCB) that may offer useful genes.

Asexual Reproduction

Crops such as maize reproduce sexually each year. This randomizes which genes get propagated to the next generation, meaning that desirable traits can be lost. To maintain a high-quality crop, some farmers purchase seeds every year. Typically, the seed company maintains two inbred varieties, and crosses them into a hybrid strain that is then sold. Related plants like sorghum and gamma grass are able to perform apomixis, a form of asexual reproduction that keeps the plant's DNA intact. This trait is apparently controlled by a single dominant gene, but traditional breeding has been unsuccessful in creating asexually-reproducing maize. Genetic engineering offers another route to this goal. Successful modification would allow farmers to replant harvested seeds that retain desirable traits, rather than relying on purchased seed.

Crops

As of 2010 food species for which a genetically modified version is being commercially grown (percent modified in the table below are mostly 2009/2010 data) include:

Crop	Traits	Modification	Percent modified in US	Percent modified in world
Alfalfa	Tolerance of glyphosate or glufosinate	Genes added	Planted in the US from 2005–2007; 2007–2010 court injunction; 2011 approved for sale	
Apples	Delayed browning	Genes added for reduced polyphenol oxidase (PPO) production from other apples	2015 approved for sale	
Canola/ Rapeseed	Tolerance of glyphosate or glufosinate High laurate canola, Oleic acid canola	Genes added	87% (2005)	21%

Corn	Tolerance of herbicides glyphosate glufosinate, and 2,4-D. Insect resistance. Added enzyme, alpha amylase, that converts starch into sugar to facilitate ethanol production. Viral resistance	Genes, some from Bt, added.	Herbicide-resistant: 2013, 85% Bt: 2013, 76% Stacked: 2013, 71%	26%
Cotton (cottonseed oil)	Insect resistance	Genes, some from Bt, added	Herbicide-resistant: 2013, 82% Bt: 2013, 75% Stacked: 2013, 71%	49%
Eggplant	Insect resistance	Genes from Bt	Negligible	Negligible
Papaya (Hawaiian)	Resistance to the papaya ringspot virus.	Gene added	80%	
Potato (food)	Resistance to Colorado beetle Resistance to potato leaf roll virus and Potato virus Y Reduced acrylamide when fried and reduced bruising	Bt cry3A, coat protein from PVY "Innate" potatoes added genetic material coding for mRNA for RNA interference	0%	0%
Potato (starch)	Antibiotic resistance gene, used for selection Better starch production	Antibiotic resistance gene from bacteria Modifications to endogenous starch-producing enzymes	0%	0%
Rice	Enriched with beta-carotene (a source of vitamin A)	Genes from maize and a common soil microorganism.		
Soybeans	Tolerance of glyphosate or glufosinate Reduced saturated fats (high oleic acid); Kills susceptible insect pests Viral resistance	Herbicide resistant gene taken from bacteria added Knocked out native genes that catalyze saturation Gene for one or more Bt crystal proteins added	2014: 94%	77%
Squash (Zucchini/ Courgette)	Resistance to watermelon, cucumber and zucchini/ courgette yellow mosaic viruses	Viral coat protein genes	13% (figure is from 2005)	
Sugar beet	Tolerance of glyphosate, glufosinate	Genes added	95% (2010); regulated 2011; deregulated 2012	9%

Sugarcane	Pesticide tolerance High sucrose content.	Genes added		
Sweet peppers	Resistance to cucumber mosaic virus	Viral coat protein genes		Small quantities grown in China
Tomatoes	Suppression of the enzyme polygalacturonase (PG), retarding fruit softening after harvesting, while at the same time retaining both the natural color and flavor of the fruit	Antisense gene of the gene responsible for PG enzyme production added	Taken off the market due to commercial failure.	Small quantities grown in China

Development

The number of USDA-approved field releases for testing grew from 4 in 1985 to 1,194 in 2002 and averaged around 800 per year thereafter. The number of sites per release and the number of gene constructs (ways that the gene of interest is packaged together with other elements)—have rapidly increased since 2005. Releases with agronomic properties (such as drought resistance) jumped from 1,043 in 2005 to 5,190 in 2013. As of September 2013, about 7,800 releases had been ap-proved for corn, more than 2,200 for soybeans, more than 1,100 for cotton, and about 900 for po-tatoes. Releases were approved for herbicide tolerance (6,772 releases), insect resistance (4,809), product quality such as flavor or nutrition (4,896), agronomic properties like drought resistance (5,190), and virus/fungal resistance (2,616). The institutions with the most authorized field re-leases include Monsanto with 6,782, Pioneer/DuPont with 1,405, Syngenta with 565, and USDA's Agricultural Research Service with 370. As of September 2013 USDA had received proposals for releasing GM rice, squash, plum, rose, tobacco, flax and chicory.

Farming Practices

Bt Resistance

Constant exposure to a toxin creates evolutionary pressure for pests resistant to that toxin. Over-reliance on glyphosate and a reduction in the diversity of weed management practices allowed the spread of glyphosate resistance in 14 weed species/biotypes in the US.

One method of reducing resistance is the creation of refuges to allow nonresistant organisms to survive and maintain a susceptible population.

To reduce resistance to Bt crops, the 1996 commercialization of transgenic cotton and maize came with a management strategy to prevent insects from becoming resistant. Insect resistance management plans are mandatory for Bt crops. The aim is to encourage a large population of pests so that any (recessive) resistance genes are diluted within the population. Resistance lowers evolutionary fitness in the absence of the stressor (Bt). In refuges, non-resistant strains outcompete resistant ones.

With sufficiently high levels of transgene expression, nearly all of the heterozygotes (S/s), i.e., the largest segment of the pest population carrying a resistance allele, will be killed before maturation, thus preventing transmission of the resistance gene to their progeny. Refuges (i. e., fields of non-transgenic plants) adjacent to transgenic fields increases the likelihood that homozygous resistant (s/s) individuals and any surviving heterozygotes will mate with susceptible (S/S) individuals from the refuge, instead of with other individuals carrying the resistance allele. As a result, the resistance gene frequency in the population remains lower.

Complicating factors can affect the success of the high-dose/refuge strategy. For example, if the temperature is not ideal, thermal stress can lower Bt toxin production and leave the plant more susceptible. More importantly, reduced late-season expression has been documented, possibly resulting from DNA methylation of the promoter. The success of the high-dose/refuge strategy has successfully maintained the value of Bt crops. This success has depended on factors independent of management strategy, including low initial resistance allele frequencies, fitness costs associated with resistance, and the abundance of non-Bt host plants outside the refuges.

Companies that produce Bt seed are introducing strains with multiple Bt proteins. Monsanto did this with Bt cotton in India, where the product was rapidly adopted. Monsanto has also; in an attempt to simplify the process of implementing refuges in fields to comply with Insect Resistance Management(IRM) policies and prevent irresponsible planting practices; begun marketing seed bags with a set proportion of refuge (non-transgenic) seeds mixed in with the Bt seeds being sold. Coined "Refuge-In-a-Bag" (RIB), this practice is intended to increase farmer compliance with refuge requirements and reduce additional labor needed at planting from having separate Bt and refuge seed bags on hand. This strategy is likely to reduce the likelihood of Bt-resistance occurring for corn rootworm, but may increase the risk of resistance for lepidopteran corn pests, such as European corn borer. Increased concerns for resistance with seed mixtures include partially resistant larvae on a Bt plant being able to move to a susceptible plant to survive or cross pollination of refuge pollen on to Bt plants that can lower the amount of Bt expressed in kernels for ear feeding insects.

Herbicide Resistance

Best management practices (BMPs) to control weeds may help delay resistance. BMPs include applying multiple herbicides with different modes of action, rotating crops, planting weed-free seed, scouting fields routinely, cleaning equipment to reduce the transmission of weeds to other fields, and maintaining field borders. The most widely planted GMOs are designed to tolerate herbicides. By 2006 some weed populations had evolved to tolerate some of the same herbicides. Palmer amaranth is a weed that competes with cotton. A native of the southwestern US, it traveled east and was first found resistant to glyphosate in 2006, less than 10 years after GM cotton was introduced.

Plant Protection

Farmers generally use less insecticide when they plant Bt-resistant crops. Insecticide use on corn farms declined from 0.21 pound per planted acre in 1995 to 0.02 pound in 2010. This is consistent with the decline in European corn borer populations as a direct result of Bt corn and cotton. The establishment of minimum refuge requirements helped delay the evolution of Bt resistance. However resistance appears to be developing to some Bt traits in some areas.

Tillage

By leaving at least 30% of crop residue on the soil surface from harvest through planting, conservation tillage reduces soil erosion from wind and water, increases water retention, and reduces soil degradation as well as water and chemical runoff. In addition, conservation tillage reduces the carbon footprint of agriculture. A 2014 review covering 12 states from 1996 to 2006, found that a 1% increase in herbicde-tolerant (HT) soybean adoption leads to a 0.21% increase in conservation tillage and a 0.3% decrease in quality-adjusted herbicide use.

Regulation

The regulation of genetic engineering concerns the approaches taken by governments to assess and manage the risks associated with the development and release of genetically modified crops. There are differences in the regulation of GM crops between countries, with some of the most marked differences occurring between the USA and Europe. Regulation varies in a given country depending on the intended use of each product. For example, a crop not intended for food use is generally not reviewed by authorities responsible for food safety.

Production

In 2013, GM crops were planted in 27 countries; 19 were developing countries and 8 were developed countries. 2013 was the second year in which developing countries grew a majority (54%) of the total GM harvest. 18 million farmers grew GM crops; around 90% were small-holding farmers in developing countries.

Country	2013– GM planted area (million hectares)	Biotech crops
USA	70.1	Maize, Soybean, Cotton, Canola, Sugarbeet, Alfalfa, Papaya, Squash
Brazil	40.3	Soybean, Maize, Cotton
Argentina	24.4	Soybean, Maize, Cotton
India	11.0	Cotton
Canada	10.8	Canola, Maize, Soybean, Sugarbeet
Total	175.2	----

The United States Department of Agriculture (USDA) reports every year on the total area of GMO varieties planted in the United States. According to National Agricultural Statistics Service, the states published in these tables represent 81–86 percent of all corn planted area, 88–90 percent of all soybean planted area, and 81–93 percent of all upland cotton planted area (depending on the year).

Global estimates are produced by the International Service for the Acquisition of Agri-biotech Applications (ISAAA) and can be found in their annual reports, "Global Status of Commercialized Transgenic Crops".

Farmers have widely adopted GM technology. Between 1996 and 2013, the total surface area of land cultivated with GM crops increased by a factor of 100, from 17,000 square kilometers (4,200,000 acres) to 1,750,000 km² (432 million acres). 10% of the world's arable land was planted with GM crops in 2010. As of 2011, 11 different transgenic crops were grown commercially on 395 million acres (160 million hectares) in 29 countries such as the USA, Brazil, Argentina, India, Canada, China, Paraguay, Pakistan, South Africa, Uruguay, Bolivia, Australia, Philippines, Myanmar, Burkina Faso, Mexico and Spain. One of the key reasons for this widespread adoption is the perceived economic benefit the technology brings to farmers. For example, the system of planting glyphosate-resistant seed and then applying glyphosate once plants emerged provided farmers with the opportunity to dramatically increase the yield from a given plot of land, since this allowed them to plant rows closer together. Without it, farmers had to plant rows far enough apart to control post-emergent weeds with mechanical tillage. Likewise, using Bt seeds means that farmers do not have to purchase insecticides, and then invest time, fuel, and equipment in applying them. However critics have disputed whether yields are higher and whether chemical use is less, with GM crops.

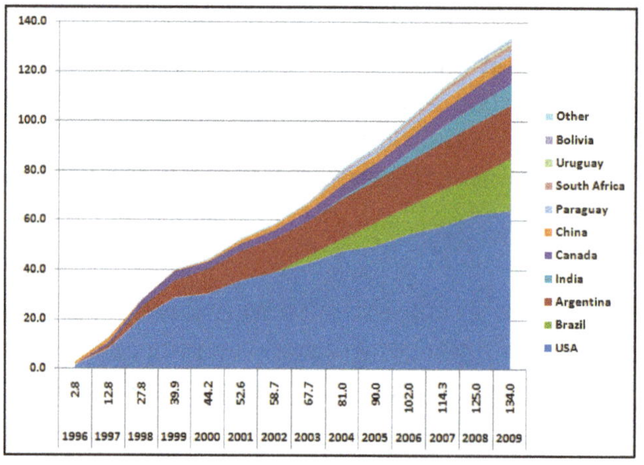

Land area used for genetically modified crops by country (1996–2009), in millions of hectares. In 2011, the land area used was 160 million hectares, or 1.6 million square kilometers.

In the US, by 2014, 94% of the planted area of soybeans, 96% of cotton and 93% of corn were genetically modified varieties. Genetically modified soybeans carried herbicide-tolerant traits only, but maize and cotton carried both herbicide tolerance and insect protection traits (the latter largely Bt protein). These constitute "input-traits" that are aimed to financially benefit the producers, but may have indirect environmental benefits and cost benefits to consumers. The Grocery Manufacturers of America estimated in 2003 that 70–75% of all processed foods in the U.S. contained a GM ingredient.

Europe grows relatively few genetically engineered crops with the exception of Spain, where one fifth of maize is genetically engineered, and smaller amounts in five other countries. The EU had a 'de facto' ban on the approval of new GM crops, from 1999 until 2004. GM crops are now regulated by the EU. In 2015, genetically engineered crops are banned in 38 countries worldwide, 19 of them in Europe. Developing countries grew 54 percent of genetically engineered crops in 2013.

In recent years GM crops expanded rapidly in developing countries. In 2013 approximately 18 million farmers grew 54% of worldwide GM crops in developing countries. 2013's largest increase was in Brazil (403,000 km² versus 368,000 km² in 2012). GM cotton began growing in India in 2002, reaching 110,000 km² in 2013.

According to the 2013 ISAAA brief: "...a total of 36 countries (35 + EU-28) have granted regulatory approvals for biotech crops for food and/or feed use and for environmental release or planting since

1994... a total of 2,833 regulatory approvals involving 27 GM crops and 336 GM events (NB: an "event" is a specific genetic modification in a specific species) have been issued by authorities, of which 1,321 are for food use (direct use or processing), 918 for feed use (direct use or processing) and 599 for environmental release or planting. Japan has the largest number (198), followed by the U.S.A. (165, not including "stacked" events), Canada (146), Mexico (131), South Korea (103), Australia (93), New Zealand (83), European Union (71 including approvals that have expired or under renewal process), Philippines (68), Taiwan (65), Colombia (59), China (55) and South Africa (52). Maize has the largest number (130 events in 27 countries), followed by cotton (49 events in 22 countries), potato (31 events in 10 countries), canola (30 events in 12 countries) and soybean (27 events in 26 countries).

Controversy

GM foods are controversial and the subject of protests, vandalism, referenda, legislation, court action and scientific disputes. The controversies involve consumers, biotechnology companies, governmental regulators, non-governmental organizations and scientists. The key areas are whether GM food should be labeled, the role of government regulators, the effect of GM crops on health and the environment, the effects of pesticide use and resistance, the impact on farmers, and their roles in feeding the world and energy production.

There is a scientific consensus that currently available food derived from GM crops poses no greater risk to human health than conventional food, but that each GM food needs to be tested on a case-by-case basis before introduction. Nonetheless, members of the public are much less likely than scientists to perceive GM foods as safe. The legal and regulatory status of GM foods varies by country, with some nations banning or restricting them, and others permitting them with widely differing degrees of regulation.

No reports of ill effects have been documented in the human population from GM food. Although GMO labeling is required in many countries, the United States Food and Drug Administration does not require labeling, nor does it recognize a distinction between approved GMO and non-GMO foods.

Advocacy groups such as Center for Food Safety, Union of Concerned Scientists, Greenpeace and the World Wildlife Fund claim that risks related to GM food have not been adequately examined and managed, that GMOs are not sufficiently tested and should be labelled, and that regulatory authorities and scientific bodies are too closely tied to industry. Some studies have claimed that genetically modified crops can cause harm; a 2016 review that reanalyzed the data from six of these studies found that their statistical methodologies were flawed and did not demonstrate harm, and said that conclusions about GMO crop safety should be drawn from "the totality of the evidence... instead of far-fetched evidence from single studies".

Advanced ways to Genetically Modify Crops

Gene Gun

A gene gun or a biolistic particle delivery system, originally designed for plant transformation, is a device for injecting cells with genetic information; the inserted genetic material are termed transgenes. The payload is an elemental particle of a heavy metal coated with plasmid DNA. This technique is often simply referred to as bioballistics or biolistics.

PDS-1000/He Particle Delivery System.

This device is able to transform almost any type of cell, including plants, and is not limited to genetic material of the nucleus: it can also transform organelles, including plastids.

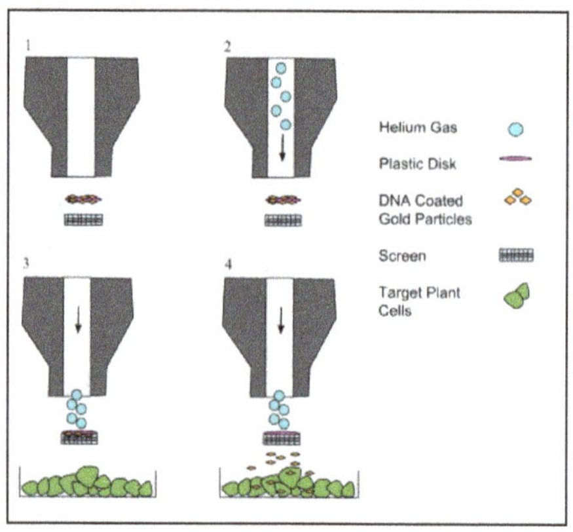

A gene gun is used for injecting cells with genetic information, it is also known as biolistic particle delivery system. Gene guns can be used effectively on most cells but are mainly used on plant cells. Step 1 The gene gun apparatus is ready to fire. Step 2 When the gun is turned on and the helium flows through. Step 3 The helium moving the disk with DNA coated particles toward the screen. Step 4 The helium having pushed the particles moving through the screen and moving to the target cells to transform the cells.

Gene Gun Design

The gene gun was originally a Crosman air pistol modified to fire dense tungsten particles. It was invented by John C Sanford, Ed Wolf and Nelson Allen at Cornell University, and Ted Klein of DuPont, between 1983 and 1986. The original target was onions (chosen for their large cell size) and it was used to deliver particles coated with a marker gene. Genetic transformation was then proven when the onion tissue expressed the gene.

The earliest custom manufactured gene guns (fabricated by Nelson Allen) used a 22 caliber nail gun cartridge to propel an extruded polyethylene cylinder (bullet) down a 22 cal. Douglas barrel. A droplet of the tungsten powder and genetic material was placed on the bullet and shot down the barrel at a lexan "stopping" disk with a petri dish below. The bullet welded to the disk and the genetic information blasted into the sample in the dish with a doughnut effect (devastation in the middle, a ring of good transformation and little around the edge). The gun was connected to a vacuum pump and was under vacuum while firing. The early design was put into limited production by a Rumsey-Loomis (a local machine shop then at Mecklenburg Rd in Ithaca, NY, USA). Later the design was refined by removing the "surge tank" and changing to nonexplosive propellants. DuPont added a plastic extrusion to the exterior to visually improve the machine for mass production to the scientific community. Biorad contracted with Dupont to manufacture and distribute the device. Improvements include the use of helium propellant and a multi-disk-collision delivery mechanism. Other heavy metals such as gold and silver are also used. Gold may be favored because it has better uniformity than tungsten and tungsten can be toxic to cells, but its use may be limited due to availability and cost.

Biolistic Construct Design

A construct is a piece of DNA inserted into the target's genome, including parts that are intended to be removed later. All biolistic transformations require a construct to proceed and while there is great variation among biolistic constructs, they can be broadly sorted into two categories: those which are designed to transform eukaryotic nuclei, and those designed to transform prokaryotic-type genomes such as mitochondria, plasmids or plastids.

Those meant to transform prokaryotic genomes generally have the gene or genes of interest, at least one promoter and terminator sequence, and a reporter gene; which is a gene used to ease detection or removal of those cells which didn't integrate the construct into their DNA. These genes may each have their own promoter and terminator, or be grouped to produce multiple gene products from one transcript, in which case binding sites for translational machinery should be placed between each to ensure maximum translational efficiency. In any case the entire construct is flanked by regions called border sequences which are similar in sequence to locations within the genome, this allows the construct to target itself to a specific point in the existing genome.

Constructs meant for integration into a eukaryotic nucleus follow a similar pattern except that: the construct contains no border sequences because the sequence rearrangement that prokaryotic constructs rely on rarely occurs in eukaryotes; and each gene contained within the construct must be expressed by its own copy of a promoter and terminator sequence.

Though the above designs are generally followed, there are exceptions. For example, the construct might include a Cre-Lox system to selectively remove inserted genes; or a prokaryotic construct may insert itself downstream of a promoter, allowing the inserted genes to be governed by a promoter already in place and eliminating the need for one to be included in the construct.

Application

Gene guns are so far mostly applied for plant cells. However, there is much potential use in humans and other animals as well.

Plants

The target of a gene gun is often a callus of undifferentiated plant cells growing on gel medium in a Petri dish. After the gold particles have impacted the dish, the gel and callus are largely disrupted.

However, some cells were not obliterated in the impact, and have successfully enveloped a DNA coated gold particle, whose DNA eventually migrates to and integrates into a plant chromosome.

Cells from the entire Petri dish can be re-collected and selected for successful integration and expression of new DNA using modern biochemical techniques, such as a using a tandem selectable gene and northern blots.

Selected single cells from the callus can be treated with a series of plant hormones, such as auxins and gibberellins, and each may divide and differentiate into the organized, specialized, tissue cells of an entire plant. This capability of total re-generation is called totipotency. The new plant that originated from a successfully shot cell may have new genetic (heritable) traits.

The use of the gene gun may be contrasted with the use of *Agrobacterium tumefaciens* and its Ti plasmid to insert genetic information into plant cells.

Humans and Other Animals

Gene guns have also been used to deliver DNA vaccines.

The delivery of plasmids into rat neurons through the use of a gene gun, specifically DRG neurons, is also used as a pharmacological precursor in studying the effects of neurodegenerative diseases such as Alzheimer's disease.

The gene gun has become a common tool for labeling subsets of cells in cultured tissue. In addition to being able to transfect cells with DNA plasmids coding for fluorescent proteins, the gene gun can be adapted to deliver a wide variety of vital dyes to cells.

Gene gun bombardment has also been used to transform *Caenorhabditis elegans*, as an alternative to microinjection.

Advantages

Biolistics has proven to be a versatile method of genetic modification and it is generally preferred to engineer transformation-resistant crops, such as cereals. Notably, *Bt* maize is a product of biolistics. Plastid transformation has also seen great success with particle bombardment when compared to other current techniques, such as *Agrobacterium* mediated transformation, which have difficulty targeting the vector to and stably expressing in the chloroplast. In addition, there are no reports of a chloroplast silencing a transgene inserted with a gene gun. Additionally, with only one firing of a gene gun, a skilled technician can generate two transformed organisms. This technology has even allowed for modification of specific tissues *in situ*, although this is likely to damage large numbers of cells and transform only some, rather than all, cells of the tissue.

Limitations

However, biolistics introduces the construct randomly into the target cells. Thus the altered DNA sequences may be transformed into whatever genomes are present in the cell, be they nuclear, mitochondrial, plasmid or any others, in any combination, though proper construct design may mitigate this. Another issue is that the gene inserted may be overexpressed when the construct is inserted multiple times in either the same or different locations of the genome. This is due to the ability of the constructs to give and take genetic information from other constructs, causing some to carry no transgene and others to carry multiple copies; the number of copies inserted depends on both how many copies of the transgene an inserted construct has, and how many were inserted. Also, because eukaryotic constructs rely on illegitimate recombination, a process by which the transgene is integrated into the genome without similar genetic sequences, and not homologous recombination, which inserts at similar sequences, they cannot be targeted to specific locations within the genome.

Agrobacterium

Agrobacterium is a genus of Gram-negative bacteria established by H. J. Conn that uses horizontal gene transfer to cause tumors in plants. *Agrobacterium tumefaciens* is the most commonly studied species in this genus. *Agrobacterium* is well known for its ability to transfer DNA between itself and plants, and for this reason it has become an important tool for genetic engineering.

The *Agrobacterium* genus is quite heterogeneous. Recent taxonomic studies have reclassified all of the *Agrobacterium* species into new genera, such as *Ahrensia*, *Pseudorhodobacter*, *Ruegeria*, and *Stappia*, but most species have been controversially reclassified as *Rhizobium* species.

Plant Pathogen

The large growths on these roots are galls induced by *Agrobacterium* sp.

A. tumefaciens causes crown-gall disease in plants. The disease is characterised by a tumour-like growth or gall on the infected plant, often at the junction between the root and the shoot. Tumors are incited by the conjugative transfer of a DNA segment (T-DNA) from the bacterial tumour-inducing (Ti) plasmid. The closely related species, *A. rhizogenes*, induces root tumors, and carries the distinct Ri (root-inducing) plasmid. Although the taxonomy of *Agrobacterium* is currently under revision it can be generalised that 3 biovars exist within the genus, *A. tumefaciens*, *A. rhizogenes*, and *A. vitis*. Strains within *A. tumefaciens* and *A. rhizogenes* are known to be able to harbour either a Ti or Ri-plasmid, whilst strains of *A. vitis*, generally restricted to grapevines, can harbour a Ti-plasmid. Non-*Agrobacterium* strains have been isolated from environmental samples which harbour a Ri-plasmid whilst laboratory studies have shown that non-*Agrobacterium* strains can also harbour a Ti-plasmid. Some environmental strains of *Agrobacterium* possess neither a Ti nor Ri-plasmid. These strains are avirulent.

The plasmid T-DNA is integrated semi-randomly into the genome of the host cell, and the tumor morphology genes on the T-DNA are expressed, causing the formation of a gall. The T-DNA carries genes for the biosynthetic enzymes for the production of unusual amino acids, typically octopine or nopaline. It also carries genes for the biosynthesis of the plant hormones, auxin and cytokinins, and for the biosynthesis of opines, providing a carbon and nitrogen source for the bacteria that most other micro-organisms can't use, giving *Agrobacterium* a selective advantage. By altering the hormone balance in the plant cell, the division of those cells cannot be controlled by the plant, and tumors form. The ratio of auxin to cytokinin produced by the tumor genes determines the morphology of the tumor (root-like, disorganized or shoot-like).

Agrobacterium in Humans

Although generally seen as an infection in plants, *Agrobacterium* can be responsible for opportunistic infections in humans with weakened immune systems, but has not been shown to be a primary pathogen in otherwise healthy individuals. One of the earliest associations of human disease caused by *Agrobacterium radiobacter* was reported by Dr. J. R. Cain in Scotland (1988). A later study suggested that *Agrobacterium* attaches to and genetically transforms several types of human cells by integrating its T-DNA into the human cell genome. The study was conducted using cultured human tissue and did not draw any conclusions regarding related biological activity in nature.

Uses in Biotechnology

The ability of *Agrobacterium* to transfer genes to plants and fungi is used in biotechnology, in particular, genetic engineering for plant improvement. A modified Ti or Ri plasmid can be used. The plasmid is 'disarmed' by deletion of the tumor inducing genes; the only essential parts of the T-DNA are its two small (25 base pair) border repeats, at least one of which is needed for plant transformation. The genes to be introduced into the plant are cloned into a plant transformation vector that contains the T-DNA region of the disarmed plasmid, together with a selectable marker (such as antibiotic resistance) to enable selection for plants that have been successfully transformed. Plants are grown on media containing antibiotic following transformation, and those that do not have the T-DNA integrated into their genome will die. An alternative method is agroinfiltration.

Transformation with *Agrobacterium* can be achieved in two ways. Protoplasts or alternatively leaf-discs can be incubated with the *Agrobacterium* and whole plants regenerated using plant tissue culture. A common transformation protocol for *Arabidopsis* is the floral-dip method: inflorescence are dipped in a suspension of *Agrobacterium*, and the bacterium transforms the germline cells that make the female gametes. The seeds can then be screened for antibiotic resistance (or another marker of interest), and plants that have not integrated the plasmid DNA will die when exposed to the correct condition of antibiotic.

Agrobacterium does not infect all plant species, but there are several other effective techniques for plant transformation including the gene gun.

Agrobacterium is listed as being the vector of genetic material that was transferred to these USA GMOs:

- Soybean.
- Cotton.
- Corn.
- Sugar Beet.

- Alfalfa.
- Wheat.
- Rapeseed Oil (Canola).
- Creeping bentgrass (for animal feed).
- Rice (Golden Rice).

Genomics

The sequencing of the genomes of several species of *Agrobacterium* has permitted the study of the evolutionary history of these organisms and has provided information on the genes and systems involved in pathogenesis, biological control and symbiosis. One important finding is the possibility that chromosomes are evolving from plasmids in many of these bacteria. Another discovery is that the diverse chromosomal structures in this group appear to be capable of supporting both symbiotic and pathogenic lifestyles. The availability of the genome sequences of *Agrobacterium* species will continue to increase, resulting in substantial insights into the function and evolutionary history of this group of plant-associated microbes.

History

Marc Van Montagu and Jozef Schell at the University of Ghent (Belgium) discovered the gene transfer mechanism between *Agrobacterium* and plants, which resulted in the development of methods to alter *Agrobacterium* into an efficient delivery system for gene engineering in plants. A team of researchers led by Dr Mary-Dell Chilton were the first to demonstrate that the virulence genes could be removed without adversely affecting the ability of *Agrobacterium* to insert its own DNA into the plant genome (1983).

List of Genetically Modified Crops

Amflora Potato

Amflora (also known as EH92-527-1) is a genetically modified potato cultivar developed by BASF Plant Science. "Amflora" potato plants produce pure amylopectin starch that is processed to waxy potato starch. It was approved for industrial applications in the European Union on 2 March 2010 by the European Commission. In January 2012, the potato was withdrawn from the market in the EU.

History

Originally registered on 5 August 1996, Amflora was developed by geneticist Lennart Erjefält and agronomist Jüri Känno of Svalöf Weibull AB.

After the European Commission's approval of the potato, BASF announced it was going to produce Amflora seed starting in April 2010 in Germany's Western Pomerania (20 ha) and Sweden (80 ha). It also announced it was planting 150 ha in the Czech Republic "for commercial aims with an unnamed partner."

Due to lack of acceptance of GM crops in Europe, BASF Plant Science decided in January 2012 to stop its commercialization activities in Europe and would no longer sell Amflora there, but it would continue seeking regulatory approval for its products in the Americas and Asia.

In 2013, an EU court annulled the approval of BASF's Amflora, saying that the EU Commission broke rules when it approved the potato in 2010.

Biology

Waxy potato varieties produce two main kinds of potato starch, amylose and amylopectin, the latter of which is most industrially useful. The Amflora potato has been modified to contain antisense against the enzyme that drives synthesis of amylose, namely granule bound starch synthase. This resulting potato almost exclusively produces amylopectin, and thus is more useful for the starch industry.

Industrial Applications

Regular potato starch contains two constituent types of molecules: amylopectin (80 percent), which is more useful as a polymer for industry, and amylose (20 percent) which often creates problems as starch retrogradation, so must be modified with chemical reactions which can be costly.

After two decades of research efforts, BASF's biotechnologists using genetic engineering succeeded in creating a potato, named "Amflora", where the gene responsible for the synthesis of amylose had been turned off, thus the potato is unable to synthesize the less desirable amylose.

Amflora potatoes would be processed and sold as starch to industries that prefer waxy potato starch with only amylopectin. Amflora is intended only for industrial applications such as papermaking and other technical applications. Europe produces more than two million metric tons of natural potato starch a year, and BASF with its Amflora product hoped to enter into this large market.

Other Possible Uses

According to *New York Times*, BASF has a second application pending for use of Amflora's potato pulp as animal feed.

Political Disagreements

Protests against the Amflora potato.

Various environmental organizations, such as Greenpeace, disagreed with the introduction of the Amflora genetically modified potato into the market. The lengthy approval process frustrated some supporters of the potato. A BASF scientist said to the *New York Times*, "it's hard when you see an innovative product go through the loops again and again. These decisions are not about science but about politics". After the potato was approved, the European Greens political party and the Italian agricultural minister criticized the approval. The International Peasant Movement La Via Campesina issued a press release on 8 March 2010 also criticizing the decision.

Reactions by Greek Politicians

After Amflora's licensing by the European Commission on 2 March 2010, the Coalition of the Radical Left's Member of Parliament for the A Thessalonikis prefecture, Tasos Kouvelis, asked the Greek Minister of Agriculture on 3 March 2010 to declare the production of the potato illegal in Greece,. On 4 March 2010 Panhellenic Socialist Movement's European Member of Parliament Kriton Arsenis submitted a question at Europarl asking about the consequences of Amflora.

PASOK's MP Maria Damanaki accepted the decision of the European Commission, while Greek Agriculture Minister Katerina Batzeli said the production of Amflora will not be allowed in Greece.

Licensing Procedure

Amflora could not be sold within the European Union without approval, and its licence could only be issued after voting at the Council of Ministers of the European Union with a 74 percent threshold of support. Two rounds of voting were held, first by experts in December 2006 and then by the agricultural ministers in July 2007, but both failed to reach the 74 percent threshold. Although the voting was by secret ballot, the *New York Times* reported that Amflora was supported by the agricultural ministers of Germany and Belgium, and was opposed by the agricultural ministers of Italy, Ireland, and Austria, while the agricultural ministers of France and Bulgaria abstained from voting.

After a licence was issued on 2 March 2010, BASF announced its intention to ask for approval of more varieties of genetically modified potatoes, such as the "Fortuna" potato.

Golden Rice

Golden rice (right) compared to white rice (left).

Golden rice is a variety of rice (*Oryza sativa*) produced through genetic engineering to biosynthesize beta-carotene, a precursor of vitamin A, in the edible parts of rice. It is intended to produce a fortified food to be grown and consumed in areas with a shortage of dietary vitamin A, a deficiency which is estimated to kill 670,000 children under the age of 5 each year.

Golden rice differs from its parental strain by the addition of three beta-carotene biosynthesis genes. The rice plant can naturally produce beta-carotene in its leaves, where it is involved in photosynthesis. However, the plant does not normally produce the pigment in the endosperm, where photosynthesis does not occur.

In 2005, Golden Rice 2 was announced, which produces up to 23 times more beta-carotene than the original golden rice. To receive the USDA's Recommended Dietary Allowance (RDA), it is estimated that 144 g of the high-yielding strain would have to be eaten. Bioavailability of the carotene from golden rice has been confirmed and found to be an effective source of vitamin A for humans.

Golden rice has met significant opposition from environmental and anti-globalization activists that claim that there are sustainable, long-lasting and more efficient ways to solve vitamin A deficiency that do not compromise food, nutrition and financial security. A study in the Philippines is aimed to evaluate the performance of golden rice, if it can be planted, grown and harvested like other rice varieties, and whether golden rice poses risk to human health. Data has not been released yet.

Golden Rice was one of seven winners of the 2015 Patents for Humanity Awards by the United States Patent and Trademark Office.

As of 2016, it is still in development.

History

The search for a golden rice started off as a Rockefeller Foundation initiative in 1982.

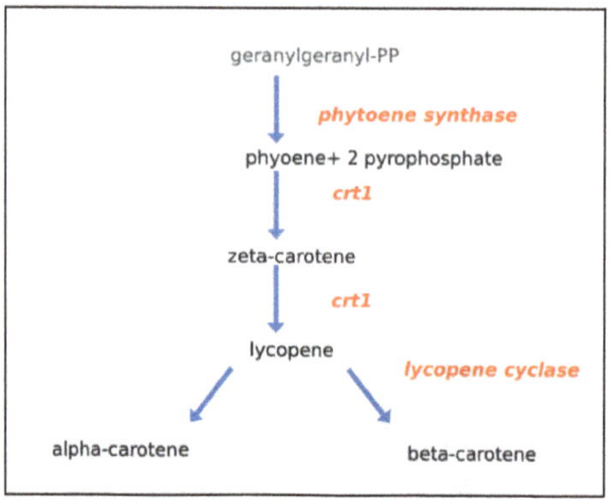

A simplified overview of the carotenoid biosynthesis pathway in golden rice. The enzymes expressed in the endosperm of golden rice, shown in red, catalyze the biosyntheis of beta-carotene from geranylgeranyl diphosphate. Beta-carotene is assumed to be converted to retinal and subsequently retinol (vitamin A) in the animal gut.

Peter Bramley discovered in 1990s that a single phytoene desaturase gene (bacterial *CrtI*) can be used to produce lycopene from phytoene in GM tomato, rather than having to introduce multiple carotene desaturases that are normally used by higher plants. Lycopene is then cyclized to beta-carotene by the endogenous cyclase in Golden Rice.

The scientific details of the rice were first published in *Science* in 2000, the product of an eight-year project by Ingo Potrykus of the Swiss Federal Institute of Technology and Peter Beyer of the

University of Freiburg. At the time of publication, golden rice was considered a significant breakthrough in biotechnology, as the researchers had engineered an entire biosynthetic pathway.

The first field trials of golden rice cultivars were conducted by Louisiana State University Agricultural Center in 2004. Additional trials have been conducted in the Philippines and Taiwan, and in Bangladesh (2015). Field testing provides an accurate measurement of nutritional value and enables feeding tests to be performed. Preliminary results from field tests have shown field-grown golden rice produces 4 to 5 times more beta-carotene than golden rice grown under greenhouse conditions.

Crossbreeding

In several countries, Golden rice has been bred with local rice cultivars. or crossbred with the American rice cultivar 'Cocodrie'. As of March 2016, golden rice has not yet been grown commercially, and backcrossing is still ongoing in current varieties to reduce yield drag.

Golden Rice 2

In 2005, a team of researchers at Syngenta produced Golden Rice 2. They combined the phytoene synthase gene from maize with *crt1* from the original golden rice. Golden rice 2 produces 23 times more carotenoids than golden rice (up to 37 μg/g), and preferentially accumulates beta-carotene (up to 31 μg/g of the 37 μg/g of carotenoids).

Genetics

Golden rice was created by transforming rice with two beta-carotene biosynthesis genes:

1. *psy* (phytoene synthase) from daffodil (*Narcissus pseudonarcissus*).
2. *crtI* (carotene desaturase) from the soil bacterium *Erwinia uredovora*.

(The insertion of a *lcy* (lycopene cyclase) gene was thought to be needed, but further research showed it is already produced in wild-type rice endosperm.)

The *psy* and *crtI* genes were transferred into the rice nuclear genome and placed under the control of an endosperm-specific promoter, so that they are only expressed in the endosperm. The exogenous *lcy* gene has a transit peptide sequence attached, so it is targeted to the plastid, where geranylgeranyl diphosphate is formed. The bacterial *crtI* gene was an important inclusion to complete the pathway, since it can catalyze multiple steps in the synthesis of carotenoids up to lycopene, while these steps require more than one enzyme in plants. The end product of the engineered pathway is lycopene, but if the plant accumulated lycopene, the rice would be red. Recent analysis has shown the plant's endogenous enzymes process the lycopene to beta-carotene in the endosperm, giving the rice the distinctive yellow color for which it is named. The original golden rice was called SGR1, and under greenhouse conditions it produced 1.6 μg/g of carotenoids.

Vitamin A Deficiency

The research that led to golden rice was conducted with the goal of helping children who suffer from vitamin A deficiency (VAD). In 2005, 190 million children and 19 million pregnant women, in 122 countries, were estimated to be affected by VAD. VAD is responsible for 1–2 million deaths,

500,000 cases of irreversible blindness and millions of cases of xerophthalmia annually. Children and pregnant women are at highest risk. Vitamin A is supplemented orally and by injection in areas where the diet is deficient in vitamin A.

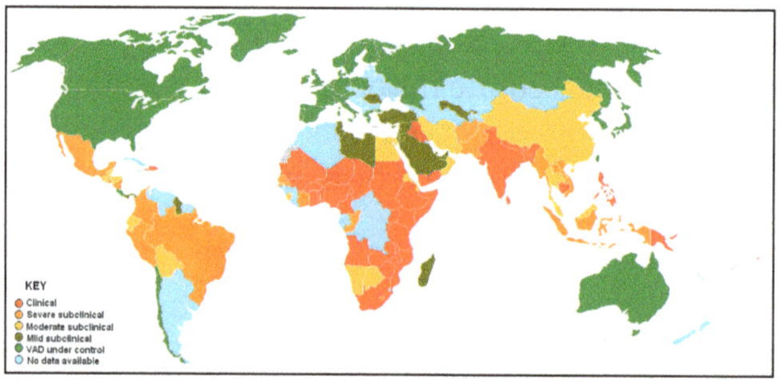

Prevalence of vitamin A deficiency. Red is most severe (clinical), green least severe. Countries not reporting data are coded blue. Data collected for a 1995 report.

As of 1999, 43 countries had vitamin A supplementation programs for children under 5; in 10 of these countries, two high dose supplements are available per year, which, according to UNICEF, could effectively eliminate VAD. However, UNICEF and a number of NGOs involved in supplementation note more frequent low-dose supplementation is preferable.

Because many children in VAD-affected countries rely on rice as a staple food, genetic modification to make rice produce the vitamin A precursor beta-carotene was seen as a simple and less expensive alternative to ongoing vitamin supplements or an increase in the consumption of green vegetables or animal products.

Initial analyses of the potential nutritional benefits of golden rice suggested consumption of golden rice would not eliminate the problems of vitamin A deficiency, but could complement other supplementation. Golden Rice 2 contains sufficient provitamin A to provide the entire dietary requirement via daily consumption of some 75g per day.

Since carotenes are hydrophobic, sufficient fat must be present in the diet for golden rice (or most other vitamin A supplements) to alleviate vitamin A deficiency. Vitamin A deficiency is usually coupled to an unbalanced diet. Moreover, this claim referred to an early cultivar of golden rice; one bowl of the latest version provides 60% of RDA for healthy children. The RDA levels advocated in developed countries are far in excess of the amounts needed to prevent blindness.

Research

Dr. José L. Domingo of the Laboratory of Toxicology and Environmental Health, School of Medicine, at Rovira i Virgili University in Spain said, "According to the information reported by the WHO, genetically modified products that are currently on the international market have all passed risk assessments conducted by national authorities." These assessments found no risk to human health. Domingo advocates continued research in the areas of GM rice and its effects on humans.

Clinical Trials/Food Safety and Nutrition Research

In 2009, results of a clinical trial of golden rice with adult volunteers from the US were published in the *American Journal of Clinical Nutrition*. The trial concluded that "beta carotene derived from golden rice is effectively converted to vitamin A in humans". A summary for the American Society for Nutrition suggested that "Golden Rice could probably supply 50% of the Recommended Dietary Allowance (RDA) of vitamin A from a very modest amount – perhaps a cup – of rice, if consumed daily. This amount is well within the consumption habits of most young children and their mothers".

It is well known that beta carotene is found and consumed in many nutritious foods eaten around the world, including fruits and vegetables. Beta carotene in food is a safe source of vitamin A.

The Food Allergy Resource and Research Program of the University of Nebraska undertook research in 2006 that showed the proteins from the new genes in golden rice showed no allergenic properties.

In August 2012, Tufts University and others published research on golden rice in the *American Journal of Clinical Nutrition* showing that the beta carotene produced by golden rice is as effective as beta carotene in oil at providing vitamin A to children. The study stated that "recruitment processes and protocol were approved". In 2015 the journal retracted the study, claiming that the researchers had acted unethically when providing Chinese children golden rice without their parents' consent.

Controversy

Critics of genetically engineered crops have raised various concerns. An early issue was that golden rice originally did not have sufficient vitamin A. This problem was solved by the development of new strains of rice. The speed at which vitamin A degrades once the rice is harvested, and how much remains after cooking are contested. However, a 2009 study concluded that golden rice is effectively converted into vitamin A in humans and a 2012 study that fed 68 children ages 6 to 8 concluded that golden rice was as good as vitamin A supplements and better than the natural beta-carotene in spinach.

Greenpeace opposes the use of any patented genetically modified organisms in agriculture and opposes the cultivation of golden rice, claiming it will open the door to more widespread use of GMOs. The International Rice Research Institute (IRRI) has emphasised the non-commercial nature of their project, stating that "None of the companies listed ... are involved in carrying out the research and development activities of IRRI or its partners in Golden Rice, and none of them will receive any royalty or payment from the marketing or selling of golden rice varieties developed by IRRI."

Vandana Shiva, an Indian anti-GMO activist, argued the problem was not the plant per se, but potential problems with poverty and loss of biodiversity. Shiva claimed these problems could be amplified by the corporate control of agriculture. By focusing on a narrow problem (vitamin A deficiency), Shiva argued, golden rice proponents were obscuring the limited availability of diverse and nutritionally adequate food. Other groups argued that a varied diet containing foods rich in beta carotene such as sweet potato, leaf vegetables and fruit would provide children with sufficient vitamin A. Keith West of Johns Hopkins Bloomberg School of Public Health has stated that foodstuffs containing vitamin A are often unavailable, only available at certain seasons, or too expensive for poor families in underdeveloped countries.

In 2008, WHO malnutrition expert Francesco Branca cited the lack of real-world studies and uncertainty about how many people will use golden rice, concluding "giving out supplements, fortifying existing foods with vitamin A, and teaching people to grow carrots or certain leafy vegetables are, for now, more promising ways to fight the problem".

In 2013, author Michael Pollan, who had critiqued the product in 2001, unimpressed by the benefits, expressed support for the continuation of the research.

Support

The Bill and Melinda Gates Foundation supports the use of genetically modified organisms in agricultural development and supports the International Rice Research Institute in developing Golden Rice.

In June 2016, 107 Nobel laureates signed a letter urging Greenpeace and its supporters to abandon their campaign against GMOs, and against Golden Rice in particular.

Protests

On August 8, 2013 an experimental plot of golden rice being developed at IRRI in the Philippines was uprooted by protesters. British author Mark Lynas reported in Slate that the vandalism was carried out by a group of activists led by the extreme left-inclined *Kilusang Magbubukid ng Pilipinas* (KMP) (unofficial translation: *Farmers' Movement of the Philippines*), to the dismay of other protesters. No local farmers participated in the uprooting; only the small number of activists damaged the golden rice crops because the farmers believe local customs which imply that killing a living rice plant is unlucky.

Distribution

Potrykus has enabled golden rice to be distributed free to subsistence farmers. Free licenses for developing countries were granted quickly due to the positive publicity that golden rice received, particularly in *Time* magazine in July 2000. Monsanto Company was one of the first companies to grant free licences.

The cutoff between humanitarian and commercial use was set at US$10,000. Therefore, as long as a farmer or subsequent user of golden rice genetics does not make more than $10,000 per year, no royalties need to be paid. In addition, farmers are permitted to keep and replant seed.

Bt cotton

Bt cotton is a genetically modified organism (GMO) cotton variety, which produces an insecticide to bollworm. It is produced by Monsanto.

Description

Strains of the bacterium *Bacillus thuringiensis* produce over 200 different Bt toxins, each harmful to different insects. Most notably, Bt toxins are insecticidal to the larvae of moths and butterflies, beetles, cotton bollworms and ghtu flies but are harmless to other forms of life. The gene coding for

Bt toxin has been inserted into cotton as a transgene, causing it to produce this natural insecticide in its tissues. In many regions, the main pests in commercial cotton are lepidopteran larvae, which are killed by the Bt protein in the genetically modified cotton they eat. This eliminates the need to use large amounts of broad-spectrum insecticides to kill lepidopteran pests (some of which have developed pyrethroid resistance). This spares natural insect predators in the farm ecology and further contributes to noninsecticide pest management.

Bt cotton is ineffective against many cotton pests such as plant bugs, stink bugs, and aphids; depending on circumstances it may be desirable to use insecticides in prevention. A 2006 study done by Cornell researchers, the Center for Chinese Agricultural Policy and the Chinese Academy of Science on Bt cotton farming in China found that after seven years these secondary pests that were normally controlled by pesticide had increased, necessitating the use of pesticides at similar levels to non-Bt cotton and causing less profit for farmers because of the extra expense of GM seeds.

Mechanism

Bt cotton was created through the addition of genes encoding toxin crystals in the Cry group of endotoxin. When insects attack and eat the cotton plant the Cry toxins are dissolved due to the high pH level of the insects stomach. The dissolved and activated Cry molecules bond to cadherin-like proteins on cells comprising the brush border molecules. The epithelium of the brush border membranes separates the body cavity from the gut whilst allowing access for nutrients. The Cry toxin molecules attach themselves to specific locations on the cadherin-like proteins present on the epithelial cells of the midge and ion channels are formed which allow the flow of potassium. Regulation of potassium concentration is essential and, if left unchecked, causes death of cells. Due to the formation of Cry ion channels sufficient regulation of potassium ions is lost and results in the death of epithelial cells. The death of such cells creates gaps in the brush border membrane.

History

Bt cotton was first approved for field trials in the United States in 1993, and first approved commercial use in the United States in 1995. Bt cotton was approved by the Chinese government in 1997.

In 2002, a joint venture between Monsanto and Mahyco introduced Bt cotton to India.

In 2011, India grew the largest GM cotton crop at 10.6 million hectares. The U.S. GM cotton crop was 4.0 million hectares, the second largest area in the world, followed by China with 3.9 million hectares and Pakistan with 2.6 million hectares. By 2014, 96% of cotton grown in the United States was genetically modified and 95% of cotton grown in India was GM. India is the largest producer of cotton, and GM cotton, as of 2014.

Advantages

Bt cotton has several advantages over non Bt cotton. The important advantages of Bt cotton are briefly:

- Increases yield of cotton due to effective control of three types of bollworms, viz. American, Spotted and Pink bollworms.

- Insects belonged to Lepidoptera (Bollworms) are sensitive to crystalline endotoxic protein produced by Bt gene which in turn protects cotton from bollworms.
- Reduction in pesticide use in the cultivation of Bt cotton in which bollworms are major pests.
- Reduction in the cost of cultivation and lower farming risks.
- Reduction in environmental pollution by the use of insecticides rarely.
- Bt cotton exhibit genetic resistance or inbuilt resistance which is a permanent type of resistance and not affected by environmental factors. Thus protects crop from bollworms.
- Bt cotton is ecofriendly and does not have adverse effect on parasites, predators, beneficial insecticides and organisms present in soil.
- It promotes multiplication of parasites and predators which help in controlling the bollworms by feeding on larvae and eggs of bollworm.
- No health hazards due to rare use of insecticides (particularly who is engaged in spraying of insecticides).
- Bt cotton are early in maturing as compared to non Bt cotton.

The main selling points of Bt cotton are the reductions in pesticides to be sprayed on a crop and the ecological benefits which stem from that. China first planted Bt cotton in 1997 specifically in response to an outbreak of cotton bollworm, *Helicoverpa armigera*, that farmers were struggling to control with conventional pesticides. Similarly in India and the US, Bt cotton initially alleviated the issues with pests whilst increasing yields and delivering higher profits for farmers.

Studies showed that the lower levels of pesticide being sprayed on the cotton crops promoted biodiversity by allowing non-target species like ladybirds, lacewings and spiders to become more abundant. Likewise it was found that integrated pest management strategies (IPM) were becoming more effective due to the lower levels of pesticide encouraging the growth of natural enemy populations.

Disadvantages

Bt cotton has some limitations:

- High cost of Bt cotton seeds as compared to non Bt cotton seeds.
- Effectiveness up to 120 days, after that the toxin producing efficiency of the Bt gene drastically reduces.
- Ineffective against sucking pests like jassids, aphids, whitefly etc.

In India

Bt cotton is supplied in India's Maharashtra state by the agri-biotechnology company Mahyco, which distributes it.

The use of Bt cotton in India has grown exponentially since its introduction in 2002. Recently India has become the number one global exporter of cotton and the second largest cotton producer in the world. India has bred Bt-cotton varieties such as *Bikaneri Nerma* and hybrids such as NHH-44.

Socio-economic surveys confirm that Bt cotton continues to deliver significant and multiple agronomic, economic, environmental and welfare benefits to Indian farmers and society including halved insecticide requirements and a doubling of yields.

India's success has been subject to scrutiny. Monsanto's seeds are expensive and lose vigour after one generation, prompting the Indian Council of Agricultural Research to develop a cheaper Bt cotton variety with seeds that could be reused. The cotton incorporated the cry1Ac gene from the soil bacterium Bacillus thuringiensis (Bt), making the cotton toxic to bollworms. This variety showed poor yield, was removed within a year, and contained a DNA sequence owned by Monsanto, prompting an investigation. In parts of India cases of acquired resistance against Bt cotton have occurred. Monsanto has admitted that the pink bollworm is resistant to first generation transgenic Bt cotton that expresses the single Bt gene (Cry1Ac).

The state of Maharashtra banned the sale and distribution of Bt cotton in 2012, to promote local Indian seeds, which demand less water, fertilizers and pesticide input, but lifted the ban in 2013.

Bt-resistant Pests

After the introduction of Bt cotton in northern China, non-target pests such as mirid bugs (*Heteroptera: Miridae*) became more abundant, because less pesticides were sprayed. In 2013, a second issue being seen across the world, was the development of Bt resistant pests limiting the usefulness of Bt crops.

Main drivers for the widespread resistance in India and China included the high proportion of Bt cotton being planted, 90% and 95% respectively in 2011, and few refuge areas.

Refuge areas of non-Bt crops limit resistance development in targeted pests. The US Environmental Protection Agency requires farmers to have refuge areas of 20%-50% non-Bt crops within 0.8 km of their Bt fields. Such requirements were not seen in China, where instead farmers relied on natural refuge areas to decrease resistance.

In 2009, a novel solution to the resistance problem was trialed in Arizona, when sterile male pink bollworms (*Pectinophora gossypiella*) were released into populations of their wild Bt-resistant counterparts. The hypothesis was that sterile males mating with the few surviving females, who had developed resistance, would lead to a decrease in pests in the following generation. There was a dramatic reduction in pink bollworms, with only two pink bollworm larvae being found by the third year of the study.

Controversies

In India, Bt cotton has been enveloped in controversies due to its supposed links with seed monopolies and farmer suicides. The link between the introduction of Bt cotton to India and a surge in farmer suicides has been refuted by other studies, with decreased farmer suicides since Bt cotton was introduced. Bt cotton accounts for 93% of cotton grown in India.

In the USA

In Hawaii, growing GMO cotton has been prohibited since 2013. Hybridization with the wild cotton species Gossypium tomentosum may be possible. Transgenic cotton is also banned in some parts of Florida.

In Africa

Burkina Faso, Africa's top cotton producer, banned GM cotton in 2016, because of economic and quality concerns.

Flavr Savr

Flavr Savr (also known as CGN-89564-2; pronounced "flavor saver"), a genetically modified tomato, was the first commercially grown genetically engineered food to be granted a license for human consumption. It was produced by the Californian company Calgene, and submitted to the U.S. Food and Drug Administration (FDA) in 1992. On May 18, 1994, the FDA completed its evaluation of the Flavr Savr tomato and the use of APH(3')II, concluding that the tomato "is as safe as tomatoes bred by conventional means" and "that the use of aminoglycoside 3'-phosphotransferase II is safe for use as a processing aid in the development of new varieties of tomato, rapeseed oil, and cotton intended for food use." It was first sold in 1994, and was only available for a few years before production ceased in 1997. Calgene made history, but mounting costs prevented the company from becoming profitable, and it was eventually acquired by Monsanto Company.

Characteristics

Through genetic engineering, Calgene hoped to slow down the ripening process of the tomato and thus prevent it from softening, while still allowing the tomato to retain its natural colour and flavor. The tomato was made more resistant to rotting by adding an antisense gene which interferes with the production of the enzyme polygalacturonase. The enzyme normally degrades pectin in the cell walls and results in the softening of fruit which makes them more susceptible to being damaged by fungal infections. Some unmodified tomatoes are picked before fully ripened and are then artificially ripened using ethylene gas which acts as a plant hormone. Picking the fruit while unripe allows for easier handling and extended shelf-life. Flavr Savr tomatoes, on the other hand, could be allowed to ripen on the vine, without compromising their shelf-life. The intended effect of slowing down the softening of Flavr Savr tomatoes would allow the vine-ripe fruits to be harvested like green tomatoes without greater damage to the tomato itself. The Flavr Savr turned out to disappoint researchers in that respect, as the antisensed PG gene had a positive effect on shelf life, but not on the fruit's firmness, so the tomatoes still had to be harvested like any other unmodified vine-ripe tomatoes. An improved flavor, later achieved through traditional breeding of Flavr Savr and better tasting varieties, would also contribute to selling Flavr Savr at a premium price at the supermarket.

The FDA stated that special labeling for these modified tomatoes was not necessary because they have the essential characteristics of non-modified tomatoes. Specifically, there was no evidence for health risks, and the nutritional content was unchanged.

The failure of the Flavr Savr has been attributed to Calgene's inexperience in the business of growing and shipping tomatoes.

Tomato Paste

In the UK, Zeneca produced a tomato paste that used technology similar to the Flavr Savr. Don Grierson was involved in the research to make the genetically modified tomato. Due to the characteristics of the tomato, it was cheaper to produce than conventional tomato paste, resulting in the product being 20% cheaper. Between 1996 and 1999, 1.8 million cans, clearly labelled as genetically engineered, were sold in Sainsbury's and Safeway. At one point the paste outsold normal tomato paste but sales fell in the autumn of 1998.

The House of Commons of the United Kingdom published a report in which they stated that the decline in sales during this period was linked to changing consumer perceptions of genetically modified crops. The report identified several possible factors, including product labeling and perception of choice, lobbying campaigns, and media attention. It concluded that the tone of media reports on the subject underwent a "fundamental shift" in response to a high profile incident in which Dr. Arpad Pusztai, a researcher for Rowett Research Institute, was fired after making a televised claim about detrimental health effects in lab rats fed a diet of genetically modified potatoes. Subsequent peer review and testimony by Dr. Pusztai led the House Science and Technology Select Committee to conclude that his initial claim was "contradicted by his own evidence." In the intervening period, Sainsbury's and Safeway both pledged that none of their house brand products would contain genetically modified ingredients.

Genetically Modified Maize

Transgenic maize containing a gene from the bacteria *Bacillus thuringiensis*.

Genetically modified maize (corn) is a genetically modified crop. Specific maize strains have been genetically engineered to express agriculturally-desirable traits, including resistance to pests and to herbicides. Maize strains with both traits are now in use in multiple countries. GM maize has also caused controversy with respect to possible health effects, impact on other insects and impact on other plants via gene flow. One strain, called Starlink, was approved only for animal feed in the US, but was found in food, leading to a series of recalls starting in 2000.

Marketed Products

Herbicide Resistant Maize

Corn varieties resistant to glyphosate herbicides were first commercialized in 1996 by Monsanto, and are known as "Roundup Ready Corn". They tolerate the use of Roundup. Bayer CropScience developed "Liberty Link Corn" that is resistant to glufosinate. Pioneer Hi-Bred has developed and markets corn hybrids with tolerance to imidazoline herbicides under the trademark "Clearfield" - though in these hybrids, the herbicide-tolerance trait was bred using tissue culture selection and the chemical mutagen ethyl methanesulfonate, not genetic engineering. Consequently, the regulatory framework governing the approval of transgenic crops does not apply for Clearfield.

As of 2011, herbicide-resistant GM corn was grown in 14 countries. By 2012, 26 varieties herbicide-resistant GM maize were authorised for import into the European Union., but such imports remain controversial. Cultivation of herbicide-resistant corn in the EU provides substantial farm-level benefits.

Insecticide-producing Corn

The European corn borer, *Ostrinia nubilalis*, destroys corn crops by burrowing into the stem, causing the plant to fall over.

Bt corn is a variant of maize that has been genetically altered to express one or more proteins from the bacterium *Bacillus thuringiensis*. The protein is poisonous to certain insect pests and is widely used in organic gardening. The European corn borer causes about a billion dollars in damage to corn crops each year.

In recent years, traits have been added to ward off Corn ear worms and root worms, the latter of which annually causes about a billion dollars in damages.

The Bt protein is expressed throughout the plant. When a vulnerable insect eats the Bt-containing plant, the protein is activated in its gut, which is alkaline. In the alkaline environment the protein partially unfolds and is cut by other proteins, forming a toxin that paralyzes the insect's digestive system and forms holes in the gut wall. The insect stops eating within a few hours and eventually starves.

In 1996, the first GM maize producing a Bt Cry protein was approved, which killed the European corn borer and related species; subsequent Bt genes were introduced that killed corn rootworm larvae.

Approved Bt genes include single and stacked (event names bracketed) configurations of: Cry1A.105 (MON89034), CryIAb (MON810), CryIF (1507), Cry2Ab (MON89034), Cry3Bb1 (MON863 and

MON88017), Cry34Ab1 (59122), Cry35Ab1 (59122), mCry3A (MIR604), and Vip3A (MIR162), in both corn and cotton. Corn genetically modified to produce VIP was first approved in the US in 2010.

Drought Resistance

In 2013 Monsanto launched the first transgenic drought tolerance trait in a line of corn hybrids called DroughtGard. The MON 87460 trait is provided by the insertion of the cspB gene from the soil microbe *Bacillus subtilis*; it was approved by the USDA in 2011 and by China in 2013.

Sweet Corn

GM sweet corn varieties include "Attribute", the brand name for insect-resistant sweet corn developed by Syngenta. and Performance Series™ insect-resistant sweet corn developed by Monsanto.

Products in Development

In 2007, South African researchers announced the production of transgenic maize resistant to maize streak virus (MSV), although it has not been released as a product.

Refuges

US Environmental Protection Agency (EPA) regulations require farmers who plant Bt corn to plant non-Bt corn nearby (called a refuge) to provide a location to harbor vulnerable pests. Typically, 20% of corn in a grower's fields must be refuge; refuge must be at least 0.5 miles from Bt corn for lepidopteran pests, and refuge for corn rootworm must at least be adjacent to a Bt field.

The theory behind these refuges is to slow the evolution of resistance to the pesticide. EPA regulations also require seed companies to train farmers how to maintain refuges, to collect data on the refuges and to report that data to the EPA. A study of these reports found that from 2003 to 2005 farmer compliance with keeping refuges was above 90%, but that by 2008 approximately 25% of Bt corn farmers did not keep refuges properly, raising concerns that resistance would develop.

Unmodified crops received most of the economic benefits of Bt corn in the US in 1996-2007, because of the overall reduction of pest populations. This reduction came because females laid eggs on modified and unmodified strains alike.

Seed bags containing both Bt and refuge seed have been approved by the EPA in the United States. These seed mixtures were marketed as "Refuge in a Bag" (RIB) to increase farmer compliance with refuge requirements and reduce additional work needed at planting from having separate Bt and refuge seed bags on hand. The EPA approved a lower percentage of refuge seed in these seed mixtures ranging from 5 to 10%. This strategy is likely to reduce the likelihood of Bt-resistance occurring for corn rootworm, but may increase the risk of resistance for lepidopteran pests, such as European corn borer. Increased concerns for resistance with seed mixtures include partially resistant larvae on a Bt plant being able to move to a susceptible plant to survive or cross pollination of refuge pollen on to Bt plants that can lower the amount of Bt expressed in kernels for ear feeding insects.

Resistance

Resistant strains of the European corn borer have developed in areas with defective or absent refuge management.

In November 2009, Monsanto scientists found the pink bollworm had become resistant to first-generation Bt cotton in parts of Gujarat, India - that generation expresses one Bt gene, *Cry1Ac*. This was the first instance of Bt resistance confirmed by Monsanto anywhere in the world. Bollworm resistance to first generation Bt cotton has been identified in the Australia, China, Spain and the United States. In 2012, a Florida field trial demonstrated that army worms were resistant to pesticide-containing GM corn produced by Dupont-Dow; armyworm resistance was first discovered in Puerto Rico in 2006, prompting Dow and DuPont to voluntarily stop selling the product on the island.

Regulation

Regulation of GM crops varies between countries, with some of the most-marked differences occurring between the USA and Europe. Regulation varies in a given country depending on intended uses.

Controversy

There is a scientific consensus that currently available food derived from GM crops poses no greater risk to human health than conventional food, but that each GM food needs to be tested on a case-by-case basis before introduction. Nonetheless, members of the public are much less likely than scientists to perceive GM foods as safe. The legal and regulatory status of GM foods varies by country, with some nations banning or restricting them, and others permitting them with widely differing degrees of regulation.

The scientific rigor of the studies regarding human health has been disputed due to alleged lack of independence and due to conflicts of interest involving governing bodies and some of those who perform and evaluate the studies.

GM crops provide a number of ecological benefits, but there are also concerns for their overuse, stalled research outside of the Bt seed industry, proper management and issues with Bt resistance arising from their misuse.

Critics have objected to GM crops on ecological, economic and health grounds. The economic issues derive from those organisms that are subject to intellectual property law, mostly patents. The first generation of GM crops lose patent protection beginning in 2015. Monsanto has claimed it will not pursue farmers who retain seeds of off-patent varieties. These controversies have led to litigation, international trade disputes, protests and to restrictive legislation in most countries.

Effects on Nontarget Insects

Critics claim that Bt proteins could target predatory and other beneficial or harmless insects as well as the targeted pest. These proteins have been used as organic sprays for insect control in France since 1938 and the USA since 1958 with no ill effects on the environment reported. While *cyt* proteins are toxic towards the insect orders Coleoptera (beetles) and Diptera (flies), *cry* proteins selectively target Lepidopterans (moths and butterflies). As a toxic mechanism, *cry* proteins bind to specific receptors on the membranes of mid-gut (epithelial) cells, resulting in rupture of those cells. Any organism that lacks the appropriate gut receptors cannot be affected by the *cry* protein, and therefore Bt. Regulatory agencies assess the potential for the transgenic plant to impact nontarget organisms before approving commercial release.

A 1999 study found that in a lab environment, pollen from Bt maize dusted onto milkweed could harm the monarch butterfly. Several groups later studied the phenomenon in both the field and the laboratory, resulting in a risk assessment that concluded that any risk posed by the corn to butterfly populations under real-world conditions was negligible. A 2002 review of the scientific literature concluded that "the commercial large-scale cultivation of current Bt–maize hybrids did not pose a significant risk to the monarch population". A 2007 review found that "nontarget invertebrates are generally more abundant in Bt cotton and Bt maize fields than in nontransgenic fields managed with insecticides. However, in comparison with insecticide-free control fields, certain nontarget taxa are less abundant in Bt fields."

Gene Flow

Gene flow is the transfer of genes and/or alleles from one species to another. Concerns focus on the interaction between GM and other maize varieties in Mexico, and of gene flow into refuges.

In 2009 the government of Mexico created a regulatory pathway for genetically modified maize, but because Mexico is the center of diversity for maize, gene flow could affect a large fraction of the world's maize strains. A 2001 report in *Nature* presented evidence that Bt maize was cross-breeding with unmodified maize in Mexico. The data in this paper was later described as originating from an artifact. *Nature* later stated, "the evidence available is not sufficient to justify the publication of the original paper". A 2005 large-scale study failed to find any evidence of contamination in Oaxaca. However, other authors also found evidence of cross-breeding between natural maize and transgenic maize.

A 2004 study found Bt protein in kernels of refuge corn.

Food

The French High Council of Biotechnologies Scientific Committee reviewed the 2009 Vendômois *et al.* study and concluded that it "..presents no admissible scientific element likely to ascribe any haematological, hepatic or renal toxicity to the three re-analysed GMOs." However, the French government applies the precautionary principle with respect to GMOs.

A review by Food Standards Australia New Zealand and others of the same study concluded that the results were due to chance alone.

A 2011 Canadian study looked at the presence of CryAb1 protein (BT toxin) in non-pregnant women, pregnant women and fetal blood. All groups had detectable levels of the protein, including 93% of pregnant women and 80% of fetuses at concentrations of 0.19 ± 0.30 and 0.04 ± 0.04 mean ± SD ng/ml, respectively. The paper did not discuss safety implications or find any health problems. The paper was found to be unconvincing by multiple authors and organizations. In a swine model,

Cry1Ab-specific antibodies were not detected in pregnant sows or their offspring and no negative effects from feeding Bt maize to pregnant sows were observed.

In January 2013, the European Food Safety Authority released all data submitted by Monsanto in relation to the 2003 authorisation of maize genetically modified for glyphosate tolerance.

Starlink Corn Recalls

StarLink contains Cry9C, which had not previously been used in a GM crop. Starlink's creator, Plant Genetic Systems had applied to the US Environmental Protection Agency (EPA) to market Starlink for use in animal feed and in human food. However, because the Cry9C protein lasts longer in the digestive system than other Bt proteins, the EPA had concerns about its allergenicity, and PGS did not provide sufficient data to prove that Cry9C was not allergenic. As a result, PGS split its application into separate permits for use in food and use in animal feed. Starlink was approved by the EPA for use in animal feed only in May 1998.

StarLink corn was subsequently found in food destined for consumption by humans in the US, Japan, and South Korea. This corn became the subject of the widely publicized Starlink corn recall, which started when Taco Bell-branded taco shells sold in supermarkets were found to contain the corn. Sales of StarLink seed were discontinued. The registration for Starlink varieties was voluntarily withdrawn by Aventis in October 2000. (Pioneer had been bought by AgrEvo which then became Aventis CropScience at the time of the incident, which was later bought by Bayer.

Fifty-one people reported adverse effects to the FDA; US Centers for Disease Control (CDC), which determined that 28 of them were possibly related to Starlink. However, the CDC studied the blood of these 28 individuals and concluded there was no evidence of hypersensitivity to the Starlink Bt protein.

A subsequent review of these tests by the Federal Insecticide, Fungicide, and Rodenticide Act Scientific Advisory Panel points out that while "the negative results decrease the probability that the Cry9C protein is the cause of allergic symptoms in the individuals examined ... in the absence of a positive control and questions regarding the sensitivity and specificity of the assay, it is not possible to assign a negative predictive value to this."

The US corn supply has been monitored for the presence of the Starlink Bt proteins since 2001.

In 2005, aid sent by the UN and the US to Central American nations also contained some StarLink corn. The nations involved, Nicaragua, Honduras, El Salvador and Guatemala refused to accept the aid.

Corporate Espionage

On December 19, 2013 six Chinese citizens were indicted in Iowa on charges of plotting to steal genetically modified seeds worth tens of millions of dollars from Monsanto and DuPont. Mo Hailong, director of international business at the Beijing Dabeinong Technology Group Co., part of the Beijing-based DBN Group, was accused of stealing trade secrets after he was found digging in an Iowa cornfield.

Genetically Modified Soybean

A genetically modified soybean is a soybean (*Glycine max*) that has had DNA introduced into it using genetic engineering techniques. In 1994 the first genetically modified soybean was introduced to the U.S. market, by Monsanto. In 2014, 90.7 million hectares of GM soy were planted worldwide, 82% of the total soy cultivation area.

Examples of Transgenic Soybeans

The genetic makeup of a soybean gives it a wide variety of uses, thus keeping it in high demand. First, manufacturers only wanted to use transgenics to be able to grow more soy at a minimal cost to meet this demand, and to fix any problems in the growing process, but they eventually found they could modify the soybean to contain healthier components, or even focus on one aspect of the soybean to produce in larger quantities. These phases became known as the first and second generation of genetically modified (GM) foods. As Peter Celec describes, "benefits of the first generation of GM foods were oriented towards the production process and companies, the second generation of GM foods offers, on contrary, various advantages and added value for the consumer", including "improved nutritional composition or even therapeutic effects."

Roundup Ready Soybean

Roundup Ready Soybeans (The first variety was also known as GTS 40-3-2 (OECD UI: MON-04032-6)) are a series of genetically engineered varieties of glyphosate-resistant soybeans produced by Monsanto.

Glyphosate kills plants by interfering with the synthesis of the essential amino acids phenylalanine, tyrosine and tryptophan. These amino acids are called "essential" because animals cannot make them; only plants and micro-organisms can make them and animals obtain them by eating plants.

Plants and microorganisms make these amino acids with an enzyme that only plants and lower organisms have, called 5-enolpyruvylshikimate-3-phosphate synthase (EPSPS). EPSPS is not present in animals, which instead obtain aromatic amino acids from their diet.

Roundup Ready Soybeans express a version of EPSPS from the CP4 strain of the bacteria, *Agrobacterium tumefaciens*, expression of which is regulated by an enhanced 35S promoter (E35S) from cauliflower mosaic virus (CaMV), a chloroplast transit peptide (CTP4) coding sequence from Petunia hybrida, and a nopaline synthase (nos 3') transcriptional termination element from Agrobacterium tumefaciens. The plasmid with EPSPS and the other genetic elements mentioned above was inserted into soybean germplasm with a gene gun by scientists at Monsanto and Asgrow. The patent on the first generation of Roundup Ready soybeans expired in March 2015.

History

First approved commercially in the United States during 1994, GTS 40-3-2 was subsequently introduced to Canada in 1995, Japan and Argentina in 1996, Uruguay in 1997, Mexico and Brazil in 1998, and South Africa in 2001.

Detection

GTS 40-3-2 can be detected using both nucleic acid and protein analysis methods.

Generic GMO Soybeans

Following expiration of Monsanto's patent on the first variety of glyphosate-resistant Roundup Ready soybeans, development began on glyphosate-resistant "generic" soybeans. The first variety, developed at the University of Arkansas Division of Agriculture, came on the market in 2015. With a slightly lower yield than newer Monsanto varieties, it costs about half as much, and seeds can be saved for subsequent years. According to its creator it is adapted to conditions in Arkansas. Several other varieties are being bred by crossing the original variety of Roundup Ready soybeans with other soybean varieties.

Stacked Traits

Monsanto developed a glyphosate-resistant soybean that also expresses Cry1Ac protein from Bacillus thuringiensis and the glyphosate-resistance gene, which completed the Brazilian regulatory process in 2010.

Genetic Modification to Improve Soybean oil

Soy has been genetically modified to improve the quality of soy oil. Soy oil has a fatty acid profile that makes it susceptible to oxidation, which makes it rancid, and this has limited its usefulness to the food industry. Genetic modifications increased the amount of oleic acid and stearic acid and decreased the amount of linolenic acid. By silencing, or knocking out, the delta 9 and delta 12 desaturases. DuPont Pioneer created a high oleic fatty acid soybean with levels of oleic acid greater than 80%, and started marketing it in 2010.

Regulation

The regulation of genetic engineering concerns the approaches taken by governments to assess and manage the risks associated with the development and release of genetically modified crops. There are differences in the regulation of GM crops between countries, with some of the most marked differences occurring between the USA and Europe. Regulation varies in a given country depending on the intended use of the products of the genetic engineering. For example, a crop not intended for food use is generally not reviewed by authorities responsible for food safety.

Controversy

There is a scientific consensus that currently available food derived from GM crops poses no greater risk to human health than conventional food, but that each GM food needs to be tested on a case-by-case basis before introduction. Nonetheless, members of the public are much less likely than scientists to perceive GM foods as safe. The legal and regulatory status of GM foods varies by country, with some nations banning or restricting them, and others permitting them with widely differing degrees of regulation.

A 2010 study found that in the United States, GM crops also provide a number of ecological benefits.

Critics have objected to GM crops on several grounds, including ecological concerns, and economic concerns raised by the fact these organisms are subject to intellectual property law. GM crops also are involved in controversies over GM food with respect to whether food produced from GM crops is safe and whether GM crops are needed to address the world's food needs.

Genetically Modified Tree

Technician checks on genetically modified peach and apple "orchards". Each dish holds experimental trees grown from lab-cultured cells to which researchers have given new genes. Source: USDA.

A genetically modified tree (GMt, GM tree, genetically engineered tree, GE tree or transgenic tree) is a tree whose DNA has been modified using genetic engineering techniques. In most cases the aim is to introduce a novel trait to the plant which does not occur naturally within the species. Examples include resistance to certain pests, diseases, environmental conditions, and herbicide tolerance, or the alteration of lignin levels in order to reduce pulping costs.

Genetically modified forest trees are not yet approved ("deregulated") for commercial use, with the exception of insect-resistant poplar trees in China. and one case of GM Eucalyptus in Brazil. Several genetically modified forest tree species are undergoing field trials for deregulation, and much of the research is being carried out by the pulp and paper industry, primarily with the intention of increasing the productivity of existing tree stock. Certain genetically modified orchard tree species have been deregulated for commercial use in the United States including the papaya and plum. The development, testing and use of GM trees remains at an early stage in comparison to GM crops.

Research

Research into genetically modified trees has been ongoing since 1988. Concerns surrounding the biosafety implications of releasing genetically modified trees into the wild have held back regulatory approval of GM forest trees. This concern is exemplified in the Convention on Biological Diversity's stance:

The Conference of the Parties, Recognising the uncertainties related to the potential environmental and socio-economic impacts, including long term and trans-boundary impacts, of genetically modified trees on global forest biological diversity, as well as on the livelihoods of indigenous and local communities, and given the absence of reliable data and of capacity in some countries to undertake risk assessments and to evaluate those potential impacts, recommends parties to take a precautionary approach when addressing the issue of genetically modified trees.

A precondition for further commercialization of GM forest trees is likely to be their complete sterility. Plantation trees remain phenotypically similar to their wild cousins in that most are the product of no more than three generations of artificial selection, therefore, the risk of transgene escape by pollination with compatible wild species is high. One of the most credible science based concerns with GM trees is their potential for wide dispersal of seed and pollen. The fact that pine pollen travels long distances is well established, moving up to 3,000 kilometers from its source. Additionally, many tree species reproduce for a long time before being harvested. In combination these factors have led some to believe that GM trees are worthy of special environmental considerations over GM crops. Ensuring sterility for GM trees has proven elusive, but efforts are being made. While tree geneticist Steve Strauss predicted that complete containment might be possible by 2020, many questions remain.

Proposed Uses

GM trees under experimental development have been modified with traits intended to provide benefit to industry, foresters or consumers. Due to high regulatory and research costs, the majority of genetically modified trees in silviculture consist of plantation trees, such as eucalyptus, poplar, and pine.

Lignin Alteration

Several companies in the pulp and paper industry are interested in utilizing GM technology to alter the lignin content of plantation trees. It is estimated that reducing lignin in plantation trees by genetic modification could reduce pulping costs by up to $15 per cubic metre. Lignin removal from wood fibres conventionally relies on costly and environmentally hazardous chemicals. By developing low-lignin GM trees it is hoped that pulping and bleaching processes will require fewer inputs, therefore, mills supplied by low-lignin GM trees may have a reduced impact on their surrounding ecosystems and communities. However, it is argued that reductions in lignin may compromise the structural integrity of the plant, thereby making it more susceptible to pathogens and disease, which could necessitate pesticide use exceeding that on traditional plantations.

Frost Tolerance

Genetic modification can allow trees to cope with abiotic stresses such that their geographic range is broadened. Freeze-tolerant GM eucalyptus trees for use in southern US plantations are currently being tested in open air sites with such an objective in mind. ArborGen, a tree biotechnology company and joint venture of pulp and paper firms Rubicon (New Zealand), MeadWestvaco (US) and International Paper (US) is leading this research. Until now the cultivation of eucalyptus has only been possible on the southern tip of Florida, freeze-tolerance would substantially extend the cultivation range northwards.

Accelerated Growth

In Brazil, field trials of fast growing GM eucalyptus are currently underway, they are set to conclude in 2015-2016 with commercialization to result. *FuturaGene*, a biotechnology company owned by Suzano, a Brazilian pulp and paper company, has been leading this research. Stanley Hirsch, chief executive of FuturaGene has stated: “Our trees grow faster and thicker. We are ahead of everyone. We have shown we can increase the yields and growth rates of trees more than anything grown by traditional breeding.” The company is looking to reduce harvest cycles from 7 to 5.5 years with 20-30% more mass than conventional eucalyptus. There is concern that such objectives may further exacerbate the negative impacts of plantation forestry. Increased water and soil nutrient demand from faster growing species may lead to irrecoverable losses in site productivity and further impinge upon neighbouring communities and ecosystems.

Disease Resistance

Ecologically motivated research into genetic modification is underway. There are ongoing schemes that aim to foster disease resistance in trees such as the American chestnut and the English elm for the purpose of their reintroduction to the wild. Specific diseases have reduced the populations of these emblematic species to the extent that they are mostly lost in the wild. Genetic modification is being pursued concurrently with traditional breeding techniques in an attempt to endow these species with disease resistance.

Current Uses

Poplars in China

In 2002 China’s State Forestry Administration approved GM poplar trees for commercial use. Subsequently 1.4 million Bt (insecticide) producing GM poplars were planted in China. They were planted both for their wood and as part of China’s ‘Green Wall’ project, which aims to impede desertification. Reports indicate that the GM poplars have spread beyond the area of original planting and that contamination of native poplars with the Bt gene is occurring. There is concern with these developments, particularly because the pesticide producing trait may impart a positive selective advantage on the poplar, allowing it a high level of invasiveness.

References

- Bashshur, Ramona (February 2013). “FDA and Regulation of GMOs”. American Bar Association. Retrieved 08, March 2020

- Final Report of the PABE research project (December 2001). "Public Perceptions of Agricultural Biotechnologies in Europe". Commission of European Communities. Retrieved 21, May 2020
- Achenbach, Joel (2016-06-30). "107 Nobel laureates sign letter blasting Greenpeace over GMOs". Washington Post. Retrieved 09, February 2020
- Staff (2009) Global Prevalence Of Vitamin A Deficiency in Populations At Risk 1995–2005 WHO Global Database on Vitamin A Deficiency. Geneva, World Health Organization, ISBN 978-92-4-159801-9, Retrieved 13, June 2020
- Hayward, M.D.; Bosemark, N.O.; Romagosa, T. (2012). Plant Breeding: Principles and Prospects. Springer Science & Business Media. p. 131. ISBN 9789401115247.
- Cotter, Jannet. "The Broken Promises of "Golden" Rice" (PDF). Greenpeace International. Greenpeace International. Retrieved 06, April 2020
- Praitis, Vida (2006). "Creation of Transgenic Lines Using Microparticle Bombardment Methods". 351: 93–108. doi:10.1385/1-59745-151-7:93. ISBN 1-59745-151-7.
- Final Report of the PABE research project (December 2001). "Public Perceptions of Agricultural Biotechnologies in Europe". Commission of European Communities. Retrieved 26, August 2020

Permissions

All chapters in this book are published with permission under the Creative Commons Attribution Share Alike License or equivalent. Every chapter published in this book has been scrutinized by our experts. Their significance has been extensively debated. The topics covered herein carry significant information for a comprehensive understanding. They may even be implemented as practical applications or may be referred to as a beginning point for further studies.

We would like to thank the editorial team for lending their expertise to make the book truly unique. They have played a crucial role in the development of this book. Without their invaluable contributions this book wouldn't have been possible. They have made vital efforts to compile up to date information on the varied aspects of this subject to make this book a valuable addition to the collection of many professionals and students.

This book was conceptualized with the vision of imparting up-to-date and integrated information in this field. To ensure the same, a matchless editorial board was set up. Every individual on the board went through rigorous rounds of assessment to prove their worth. After which they invested a large part of their time researching and compiling the most relevant data for our readers.

The editorial board has been involved in producing this book since its inception. They have spent rigorous hours researching and exploring the diverse topics which have resulted in the successful publishing of this book. They have passed on their knowledge of decades through this book. To expedite this challenging task, the publisher supported the team at every step. A small team of assistant editors was also appointed to further simplify the editing procedure and attain best results for the readers.

Apart from the editorial board, the designing team has also invested a significant amount of their time in understanding the subject and creating the most relevant covers. They scrutinized every image to scout for the most suitable representation of the subject and create an appropriate cover for the book.

The publishing team has been an ardent support to the editorial, designing and production team. Their endless efforts to recruit the best for this project, has resulted in the accomplishment of this book. They are a veteran in the field of academics and their pool of knowledge is as vast as their experience in printing. Their expertise and guidance has proved useful at every step. Their uncompromising quality standards have made this book an exceptional effort. Their encouragement from time to time has been an inspiration for everyone.

The publisher and the editorial board hope that this book will prove to be a valuable piece of knowledge for students, practitioners and scholars across the globe.

Index

www.ingramcontent.com/pod-product-compliance
Lightning Source LLC
LaVergne TN
LVHW080910210926
848485LV00004B/622